What Is:
Electro-Mechanical Packaging

By

JOHN L. BISOL

What Is: Electro-Mechanical Packaging

DISCLAIMERS:

FAIR USE NOTICE:

10 9 8 7 6 5 4 3 2 1   First Edition

Printed in the United States of America

FOREWORD

Let all who build beware
The load, the shock, the pressure
Material can bear.
So, when the buckled girder
Lets down the grinding span,
The blame of loss, or murder,
Is laid upon the man.
Not on the Stuff – the Man

– Rudyard Kipling, *Hymn of Breaking Strain*

An Electro-Mechanical Packaging Engineer is a "Hybrid" Mechanical Engineer

The E/M Engineer is able to apply a rich background knowledge of Manufacturing processes, electrical component selection, and traditional mechanical Engineering disciplines to develop high quality, and cost effective packaging solutions for all types of consumer, commercial, military or aerospace products/applications.

Electro-Mechanical Packaging is a major discipline within the field of Mechanical Engineering, and includes a wide variety of technologies. It refers to product enclosures and the unique protective features built into the product itself, and not (only) to a shipping container. (Electro-Mechanical Packaging applies both to end products and to the components.)

Electro-Mechanical Packaging of an electronic system must consider protection from mechanical damage, cooling, radio frequency noise emission, protection from electrostatic discharge, maintenance, operator convenience, and cost. Prototypes and industrial equipment made in small quantities, may use standardized commercially available enclosures such as card cages or prefabricated boxes. Mass-market consumer devices may have highly specialized packaging to increase consumer appeal. The same electronic system may be packaged as a portable device or adapted for fixed mounting in an instrument rack or permanent installation. Packaging for aerospace, marine, or military systems imposes different types of design criteria.

The Electro-Mechanical Packaging Engineer is a vital resource in product development and must balance many objectives and practical considerations when selecting packaging methods. Hazards to be protected against: mechanical damage, exposure to weather and dirt, electromagnetic interference, etc.

- Heat dissipation requirements
- Tradeoffs between tooling capital cost and per-unit cost
- Tradeoffs between time to first delivery and production rate
- Availability and capability of selected suppliers
- User interface design and convenience
- Ease of access to internal parts when required for maintenance
- Product safety, and compliance with all regulatory standards
- Aesthetics, and other Marketing considerations
- Service life and reliability

The Electro-Mechanical Packaging Engineer is (also) responsible for product reliability qualification(s) including the following types of environmental stresses:

- Burn-in
- Temperature cycling
- Thermal shock
- Solderability
- Autoclave
- Visual inspection
- Hermetic/Moisture resistance
- Other – as required by Marketplace Considerations

The Electro-Mechanical Packaging Engineer, as part of the design team, is (also) responsible to:

- Define the mechanical requirements for System/Package solutions based upon customer requirements.
- Develop new packaging and assembly process qualification programs.
- Perform mechanical integrity analysis of equipment utilizing appropriate tools (stack-up, thermal and package stress analyses).
- Perform equipment characterizations, including cost effectiveness studies.
- Act as a liaison with vendors for equipment hardware, substrate manufacturers, and other tooling vendors.

- Maintain product quality while developing and introducing equipment cost reduction programs.
- Coordinate the introduction of new equipment and assembly processes into production.
- Prepare and/or updates specifications for piece parts of integrated circuits or semiconductor assemblies.
- Manage prototype assembly lab and technician(s) for assembled product solutions: (wire-bonding, SMT and other small-scale prototype assembly operations).
- Manage layout designers and equipment Engineering group for System equipment initiatives.

INTRODUCTION

Normally, an Electro-Mechanical Packaging Engineer (initially,) has a degree in Mechanical Engineering.

Essential Job Functions of an Electro-Mechanical Packaging Engineer – (Including approximate % of Time for Each task): [1]

60% – Lead and perform Engineering design tasks related to design and manufacture of mechanical packaging elements for electronic components to be used in demanding customer environments. Review higher system level requirements and identify or derive critical mechanical design requirements. Design duties usually include: electronics packaging of complex electronics, defining electrical-mechanical interfaces, full participation in team-development environment. Provide technical oversight and technical direction for electronic packaging design tasks to insure compliance to requirements and inclusion of best practices that maximize production efficiency.

20% – Support Engineering development leadership team. Lead and participate in design peer reviews. Provide oversight for adherence to Engineering development processes and proper execution and result attainment. Collect and analyze Engineering process metric data as required and applicable. Mentor Engineering team to increase technical depth and expertise in electronic packaging design for application in harsh environments. Assess and update mechanical design development processes for electronic component packaging.

20% – Create design development planning to support product development scheduling and cost analysis. Collect and submit Engineering performance data for design tasks as required in support of Cost Accountant Manager scheduling and budgeting monitoring. Develop and present technical presentations to status cable development designs and technical execution.

> **Note:** Must be able to obtain and maintain a U.S. Security Clearance at the appropriate level (requires U.S. Citizenship).

---

[1] Obviously, the % will vary, as will the job description elements based upon the needs of the Corporation. REMEMBER: "Everything is fluid at a high temperature."

Additional Requirements and Preferred Qualifications
- Demonstrated oral and written communication skills to be able to coordinate and collaborate with local Operations Groups, as well as distant-located Engineering organizations and customers.
- Direct experience with Engineering design and production processes and disciplines.
- Able to create, specifications, read and understand Engineering drawings, electrical schematics and wiring diagrams.

Employers expect you to know the technical stuff – after all, you have completed your college education, earned your degree and now, you're ready to "design stuff" using fancy software or CAD tools.

It's quite a shock to realize that your abilities in the "soft," nontechnical skills, will be the discriminators in getting hired, retained and promoted. (Don't forget you're not the only person who can "do the math")

One of the "secrets to success" will be your ability to document your activities and more precisely, proving to management that you have the ability to quantify your design efforts in terms that can be understood. (This demonstrates to Managers that you "know" where you're going.) You will succeed in doing this, if you can create individual documents for different processes and procedures you perform, rather than one massive document that tries to explain your entire job.

Certainly, if you are part of a design team, your job is to make sure that two things happen during the design process:
- You have a complete, articulated "Roadmap" of your intended design thesis.
- Your team-members must have access to your roadmap and be able to add or subtract those criteria which will conflict with the Corporate Product Requirements Document.

When you write these documents, you must be thorough, but also concise. For example:
- Always use the approved Corporate "Word-Processing" format, including any special templates, (if available.)

- Always put a date stamp at the top of the pages, (the header) showing when the file was most recently updated.
- Maintain a "living-revision" page.
- Keep copies of "Old Revisions" in a Project Folder

You might be tempted to take several weeks to complete these types of documents. That would be a mistake – especially if your company's product development cycles are measured in months. Stay on schedule! Remember to:

- Keep the master copies of your documents in a shared location so that other people can retrieve them easily.
- Keep a local backup of the files on your desktop or in a private cloud-based system (or both).
- Review and revise your documents as needed during the design process. Think of them as living documents that are never "finished." They're ever-evolving.
- At the completion of the project: release the "final" document to the Documentation Specialists for incorporation into the Product Literature/Specifications.
- Tell your boss and colleagues you're creating and maintaining these new records, as they may have valuable input, too.
- Don't include clip art simply to jazz up the pages. (Everyone hates it and it is not professional.)
- Don't go crazy formatting the text.
  - Keep it simple.
  - Use bold type for subheadings.
  - Use bullets and numbered lists as appropriate to help people see relevant information easily.
  - Keep paragraphs very short.
- Don't be overly protective of your documents. You've created them to help the organization, not for your sole benefit.

Give yourself time to reflect on the processes/specifications and return to the documents frequently to check whether you've recorded the information you want to convey accurately. (More importantly – does that information match the Corporate philosophy, the Design Goals, and the Cost Restraints)

This "Documentation" forms the basis for the Engineering Specification that guides your design decisions. The specification focuses on your field of expertise. (You are defining "Your design

Criteria.") In general, the specification should, at least, include the following points:

- Introduction: A brief description of equipment installation location, installation type, and nature of the installation environment.
- Service Conditions: An in-depth description of the installation conditions including humidity, temperatures, winds, moisture levels and chemical contamination. It can also include a brief description of the indoor or outdoor conditions depending on usage.
- Standards: Highlight the most appropriate and relevant international norms desired. Ensure that the standards are updated and clearly listed to minimize chances of confusion during quality assurance tests. To be on the safe side, specify the standards to avoid uncertainty due to similar standards – depending on the region use both Metric Units and Imperial Units.
- Definitions: Misinterpretation is rife when sourcing electrical or mechanical equipment/components. It is risky to assume that the manufacturer or supplier is well-versed in your industry-specific jargon. Clearly define such terms and phrases in the specification, especially when it has different meanings in different standardizations.
- Performance Requirements: Clearly state the expected service life of the equipment, both lifetime, and in-between maintenance schedules.
  - Though this depends on the equipment type, it is easy for an Engineer to look up and specify the expectations of large electrical equipment such as generators, motors and such.
  - Do note that the mode of operation, as well as the rated duty for LV (low-voltage) and MV (medium-voltage) equipment, usually has a significant impact on the performance requirements; therefore, ensure you highlight this factor.
- Design Requirements: Clearly define the degree of protection required based on the area of usage. Outdoor, indoor and levels of hazards require different design and construction requirements, therefore, highlight this factor. In the same vein, underline crucial fittings such as protective devices, circuit

segregation, lifting points, and such. For small, hand-held equipment, specific "Human-Factors" will predominate.

- Inspection and Testing: The Engineer should pay particular attention to these factors, as design mistakes are costly. Ensure that you clearly state the required inspections and tests well-before anticipated product delivery. To be on the safe side, use international standards as references and acceptable limits for the best results and to save time.
- Appendices: These contain the definitions, and extra information not highlighted in the other parts of the document.

Your "actual function" in the company is (not only) to "design stuff," but to make certain that what you design, what you specify and that your documents meet three vital contingencies:

1. You must meet the "Product Design Goals" as set forth in the Product Description Documents and the expectations of Marketing and Manufacturing.
2. Your designs must be "Cost-Efficient" and meet the Product Price goals.
3. Your design must meet the Quality and Reliability goals for Manufacturing, Marketing and the Consumer.
   a. This means more than just having a "durable" design – it means that you have analyzed potential failures and spent sufficient time and development resources in those areas/components that will benefit all aspects of the product.
   b. You must balance the cost vs. reliability (interpreted as "quality" by the customer,) and produce a design at the lowest lost and the highest quality.

The role of an Electro-Mechanical Packaging Engineer is more than just "designing-stuff" using fancy software or CAD tools. The role requires a broad approach to design issues, interfacing with a variety of design team members and focusing on the customer needs.

Remember: Technical Competency can mean a lot more than just "Calculus." Understanding your value in applying your problem-solving skills to a wide array of Electro-Mechanical issues is the core of your career success!

Section 1: The Mundane

Abstract: Very often in their daily work assignments employees are asked to deal with various "design reviews" or "phase review" processes. This is done, at times, without an understanding of what is really going on in the design process. This section will overview the individual pieces of the system design procedure so that the entire design process can be visualized. Too often a "review" is based on a cost-driven analysis of the problem or some profit/loss statement in a printed business plan. Although to be successful every company should show a profit (a motherhood statement,) every company should also produce a quality product. All companies are in the _systems_ business whether it is recognized by management or not. (Definition: Whenever two elements are joined by a link it is a <u>system</u>.)

System Design

The system design process can best be described by assuming that any design passes through well-defined phases in chronological order, realizing, however that the phase is often unrecognizable until it has passed.

Phase I: Initiation Phase - The purpose of this phase, which lasts from one day to one month, is to initiate the project. The system team, at this stage, consists of one to three people. The team leader talks with everyone connected with the problem who will listen and respond. Mentally, the design process has already begun, and may be visualized as taking place in two parts (Fig. 1).

EXTERIOR	INTERIOR
Statement of Problem	Suggested Solution

The words <u>EXTERIOR</u> and <u>INTERIOR</u> are embryonic forerunners of the two essential parts of system design: that portion concerned with the requirements on the system and its environment, i.e., things outside the system; and that portion concerned with design choices relative to equipment, procedures, and people, i.e., the system itself.

<u>Reporting</u>: The product of the system-design team is always a document. That document rarely covers more than a few pages. The document has the following content:
1. A statement of the problem (primarily the result of discussions, not based on experiment or detailed analysis)
2. A suggested set of solutions (with a preference indicated for each)

3. A statement of the type and number of system-design team personnel needed and a schedule for their acquisition (this team will later lead the project)
4. A <u>rough estimate</u> of the time and money needed for completing the project

Phase II: Organization - With the project under way, the task during the organization phase, is threefold:

- to bring the system team up to strength;
- to plan for the total-project job;
- to begin the choice of a preferred solution.

This phase may last from 2 weeks to three months (for large projects,) it is over when the team has been built to strength and a smoothly functioning group has begun delineating in greater detail the system whose first hazy outlines began to form earlier.

The system team should generally include at least five people to provide the multiplicity of viewpoints required to avoid oversight of good solutions. A maximum effective team size is about 12 people, since larger groups tend to split into factions backing different ideas. System solutions are not unique, and an adequate solution should be chosen out of those which seem most favored, so that work can go forward. A reasonable amount of study should be expended on alternative methods, but when the time for decision arrives, any decision is better than no decision.

EXTERIOR	INTERIOR
A Statement of the Problem Leads to:	Single-Thread Design
A Mathematical Model that Leads to:	OR: High-Traffic Design
The Design of Experiments	OR: Competitive Design

Fig. 2. Steps in system design-organization phase

The plans for the total project should include a time schedule and growth curve for money, personnel, and materials to complete the project, although it should be understood by those concerned that the estimates might be considerable in error at this stage. The errors will usually be in the direction of underestimating requirements for time, money, and personnel unless the system-design team is very experienced. The plans should also include a good indication of requirements for field workers, study groups, and equipment groups, the last-named subdivided into groups within the main organization and subcontract groups. Some indication should usually be given of consultant services which will become desirable.

In beginning the solution of the problem, the team executes a set of design steps such as those shown in Fig. 2. The start of the problem includes the choice of a measure of effectiveness and the establishment of design criteria. A repeatedly revised problem statement reflects, in its successive versions, the state of knowledge of the required system. The mathematical model serves as an aid to understanding the environment of the system and the proposed solution:

❖ Reporting: During this phase, ideas will change rapidly, and unless adequate reporting is undertaken it will be difficult to keep all those. associated with the task abreast of new developments. Management and customer(s) are interested in progress, as is the equipment department head who may expect to be pressed into service later; members of the system team itself will be informed by reporting. Periods of about 2 weeks are most useful for reporting at this stage.

The document, which is the product of the organization phase, will briefly state the composition of the system team at the beginning, and tell the project plans at the end. The main body of the report will include a clear problem statement, with much more content than was possible earlier, and several reasonable solutions of the problem, of which one is generally chosen as most favored.

Steps in System Design
The cycle of performing the steps in system design begins in the organizational phase. These same steps, (and others,) will be repeated repeatedly as the design effort proceeds, each time with some additional refinement. The steps in exterior system design are statement of the problem elaboration of the mathematical model, design of experiments, and conduct of experiments. Later steps will include designation of subsystems and design of subsystems.

Phase III: Preliminary Design

- The purpose of the preliminary design phase is to achieve a first version of what may truly be called "the system." The preliminary design phase may last from 1 month to a year and is over when a good set of functional specifications has been written.

Enough service groups are required to carry out the experiments whose design was begun earlier, (by service groups it is meant groups of service to the system-design effort). These experiments provide data with which to fashion a better mathematical model and simultaneously provide the required insight to be gained only be experience in the field.

At the same time, equipment (component) groups have begun to look into the problem of design, with small amounts of Engineering talent assigned to the investigation of special problems.

This preliminary design phase is one of answering questions which have only equivocal answers. Many of these questions arise again and again in the design of different systems:

- Should the organization of the system be centralized, decentralized, or a combination of both?
- Is it worthwhile to standardize every input, or should provision be made for different types?
- Alternatively, even more difficult, where should standardization take place?
- Should a human function be replaced by an automatic one?
- Is presentation of information to the user best made visual or auditory?
- Is it best to buffer inputs (to the system or one of its components) and treat them on a queue basis or should multiple channels be provided?
- Alternatively, that is the same thing in communication: should a channel be made wideband or narrowband with a delay?
- Is information best passed around and left to human beings to act on (broadcast control,) or should information be addressed to each participant (party line,) or should a separate communication channel be provided for each participant (private line)?
- How long a life should a component have?
- How large is the smallest replaceable unit?

- These questions are the essence of every system design, and thorough system design is necessary because the answers are never a simple choice.

There are always qualifications, missing information, unpredictable future events – in short, the answers are always equivocal

❖ Reporting: During the preliminary-design phase, reporting should occur about once a week. Each report accounts for minor modifications in ideas and some of the detailed study and experiment. However, efforts are mainly directed at the preparation of a document which is the first version of the functional specifications of the system; when the system is sufficiently well. understood, the execution of this task will be relatively straightforward.

The early part of such a document should conduct the reader in step-by-step fashion through the reasoning that led to the choices made; while the latter part of the document contains environmental factors, etc. The body contains the following information:

1. A description of over-all system operation in considerable detail
2. A clear-cut delineation of the subsystems
3. For each subsystem:
    a. A complete statement of the form, number, and time of occurrence of its input and output
    b. A clear and complete statement of its functioning, i.e., its operation on its input to product; its output
    c. Limiting specifications concerning allowable over-all sizes, weights, etc.
    d. At least one method of physically realizing the proposed function within these limitations, plus any supplemental information concerning investigations carried on of other methods

Phase IV: Principal Design

- The purpose of the principal design phase is to refine the functional specifications. This phase may last as long as a year; it is over when a set of design specifications has been produced. These specifications are functional and are much like the output of the previous phase. They differ in two important respects:

- o They are much more detailed
- o They are frozen; i.e., the system design team has agreed that the equipment people can go ahead and build a prototype without any further changes other than those demanded by the equipment people themselves.

❖ <u>Reporting:</u> During the principal design phase, the system team should report approximately every 3 months. Reports by component groups should be made as required, depending on the nature of the problem. The product of the principal design phase is a set of design specifications, carefully fashioned so that it meets the demands set up by customer(s,) management, project management, system designer, and most important the fabricator of equipment. It is ready for the construction of a prototype, even though that construction probably began some time before.

Phase V: Prototype Construction

- The purpose of this phase is to build an Engineering prototype. The phase should last not more than 2 months for a small system, nor 1 year for a large one, and it is over when the prototype has been delivered, ready for test. The term prototype is used for a wide variety of types of construction, covering the entire span between breadboard and production model and occasionally even overlapping these. Employees distinguish here only the Engineering prototype and the production prototype.

The Engineering prototype, as delivered for test, will often be largely hand-made. It will include one complete single thread of the system, usually with provisions for simulating the additional inputs, together with critical portions of the high traffic elements of the system. For example, a time-shared computer would be supplied in its entirety, together with the all-important input output equipment (which may however comprise only a single item of multiplexed components); it would not be sufficient to test the single thread with a slower computer adequate for a single input. The prototype will also include test equipment. In a good prototype this will not be laboratory, equipment, but a prototype of the test equipment which will eventually be built into the production models of the system.

The number of personnel involved in the system team during this phase of the project need not have changed, even though large system projects may involve hundreds of people. If the design has been well done, this phase should last a relatively small portion of the total time. However, with poor designs, as is always true of cut-and-try efforts, this construction may take the better part of the project time. Too frequently the demand for hardware is irresistible, and the phase begins too early, at great waste.

During prototype construction, reporting should continue at weekly intervals. In addition, effort should be applied early to the instruction and maintenance manuals for the prototype itself. If these are not supplied when the prototype is delivered for test, many months may be wasted or, worse, the tests will be improperly performed and an unjustly adverse report delivered on the system.

Phase VI: Test, Training, and Evaluation
- As there are many types of prototypes, so there are many types of tests: Engineering test, user's tests, and the like. We shall distinguish here between the test of the Engineering prototype, which is hand-made and operated by (or under supervision of) the Engineers who built it, and "evaluation" of the production prototype, which is, (or can be,) built by usual production methods and is operated by those who will staff it in the field. Training of such operators should begin during the prototype construction phase, and these operators should be used where possible during the tests.

There are several solutions of any system problem, and almost every one of them can be made to work, with sufficient time, effort, and money.

The purpose of system test is to prove that things work as they were designed to work and to remove the inevitable "bugs." Although there is no question that the system will eventually work, there is also no question that it will not work perfectly (if indeed it works at all,) during early trials. If the system design has been properly done, then within a few weeks to a month, the system will have been proved, and the production prototype can be undertaken – in fact, it is often under-way when the prototype tests have begun. If there has been a serious fault in the system design, or if the prototype construction phase has been unduly foreshortened by demand for early test, then the test phase may go on for months; in the long run, such methods waste both time and money.

The result of system evaluation is even more of a foregone conclusion than the result of test. The purpose of system evaluation is to decide whether the overall, final, design accomplishes its objectives. (However, this may never really be achieved for a large-scale system as used in the field.) If the system design effort has been reasonably well executed, then a system which works (in the sense of test) will accomplish its objectives (in terms of evaluation). Evaluation of a large-scale system can only be performed before construction begins – using sophisticated modeling techniques. Of course, during the time lapse between initiation and evaluation, many things will have changed. The problem itself, may have altered; technology may have made unforeseen but pertinent advances; and many ideas may have arisen in the course of the design process. All these things may make it desirable to repeat the process again through another cycle. In this way, evaluation at the end of the system design process overlaps evaluation at the beginning of the system design process.

The Single Thread

The delineation of a single thread solution of a system problem is accomplished when it is possible to state exactly what will happen to every possible input at every stage of its passage through the system or to describe every response which it will evoke in the system. When completely accomplished, all possible outputs of the system will also have been stated. It is often possible to state such a solution without indicating what is to be done in case two inputs occur at the same time or two responses are evoked at the same step in the system at the same time: that is, without completely solving the high traffic problem.

The first step in arriving at a solution of this part of the problem is to draw a functional block diagram. Where a system is already in existence, the functional block diagram of this system is a good starting point.

If no present system exists (although in our experience one always has,) the functional block diagram will have to be invented. In order to fill in this diagrams it will, of course, be necessary to understand thoroughly, from exterior design, the operation of the present system and the relationship of each of the parts to the "whole."

The next step consists of modification of the functional block diagram to be in accord with both the potential solution which occurs to the Electro-Mechanical Packaging Engineer and the requirements which have been previously established by exterior system design.

Implementation of these functions is accomplished by means of equipment within the experience of the system design team, with the help of appropriate specialists (component Engineers). This results in an (actual) equipment block diagram. The next step requires detailing of the component blocks.

The products of this portion of the solution of the system are: the equipment block diagram which makes possible the delineation of inputs, outputs, and mode of operation inside of each equipment block; and the component block diagram.

Equipment Block Diagram. After the system functions have been determined, it is necessary to implement them with actual equipment. Such diagrams should be kept simple and explicit. It is not necessary to indicate all the minor connections. The major purpose of the equipment block diagram is to allow visualization of the flow of information throughout the system and a consequent clear statement of inputs, outputs, and mode of operation on inputs. Where the human being enters, it is sometimes desirable to indicate that presence in the system and to use differing designations for lines connecting equipment to equipment and for those connecting human beings with equipment. (The Human Factors Engineer uses squares for equipment and circles for human beings in such block diagrams.)

Component Block Diagram: Each piece of equipment will correspond as a complete description of the component. It is, of course, not adequate for the construction of the required element, but it is an excellent starting point for a component designer.

The analysis and test of the performance of a proposed component are best left in the hands of equipment groups, but the preparation of a specification for such analysis and test is properly the function of the system Engineer.

Choice of Subsystems: In connection with the single thread solution, it is necessary to designate the various subsystems which will make up the larger system. The choice of such subsystems depends upon many factors, a few of which are stated here. One obvious factor is the desirability of having a subsystem in one geographical location. Another is that the subsystem should have as few inputs and outputs as possible. it is particularly important that the links which are cut by subsystem boundaries are not those which cause the most trouble in interconnections.

A "Different Approach" To Design Philosophy

The following discussion presents a "Different Approach" to the design process (I know you want to "Design Stuff" using fancy software or CAD tools – but the mundane operations of the corporation are important to understand in this context: you are a member of a "Design Team" and as such, you must know what the team's strategy is to achieve a successful product design at the lowest cost to the corporation. Your "function" is not only to be able to "Design Stuff," but to articulate clearly to all team-members what your "design" will and will not do, based upon the needs of the customer and the marketplace. Your goal is to produce documentation to explain this and to use that document as your own guideline during the design phase.)

"Eight Simple Steps to Product Design" [2]

#1. Idea Generation

The development of a product will start with the concept. The rest of the process will ensure that ideas are tested for their viability, so in the beginning all ideas are good ideas. (To a certain extent!)

Ideas can, and will come, from many different directions. The best place to start is with a SWOT analysis, (Strengths, Weaknesses, Opportunities and Threats,) which incorporates current market trends. This can be used to analyze your company's position and find a direction that is in line with your business strategy.

In addition to this business-centered activity, there are methods that focus on the customer's needs and wants. This could be:
- Under-taking market research
- Listening to suggestions from your target audience – including feedback on your current products' strengths and weaknesses.
- Encouraging suggestions from employees and partners
- Looking at your competitor's successes and failures

Input: Marketing, Engineering, H/W or S/W Architects, the Company "Roadmap" (Roadmap refers to understanding the "core" business)

Action: Investigate and evaluate concepts, market opportunities, customer survey, Define & Measure

Output: Concept sketches, block diagrams, functional spec, Marketing requirements doc, product concept doc, estimates for product cost, sales volume, resources available, schedule, ROI (Return on Investment,) Voice of Customer, Critical to Quality, Statement of Work, Request for Quote.

Team: Marketing Product Mgr., H/W & S/W Mgrs., Operations PM, (Project Manager) Engineering PM, Customer Service, Manufacturing.

Y/N: Approval to proceed through next phase

---

2 Reference: http//www.business2community.com/product-management/eight-simple-steps-for-new-product-development-0560298#RB1rIhUgWD5CxagZ.99

Planning Phase:

Input: MRD, PCD, PRD, PBS

- MRD = Marketing Requirements Document
- PCD = Printed Circuit Design
- PRD = Product requirements Document
- PBS = Product Breakdown Structure

Action: Functional Plans, Project Plan, Risk Analysis & Mitigation Planning, Test Plans, Detailed P & L, (Profit & Loss) Detailed Schedule, Architectural Committee Review

Output: Product Spec, Functional Spec, Schedule, Program Budget, Program Boundaries

Team: Cross-Functional Leads or Managers, Eng. PM, Project PM

Y/N Approval: commit human resources and spend $'s on prototype

Functional Plans: (Draft Documents for review & conflict analysis)

- Hardware
- Software
- System Test
- S/W QA
- Mechanical
- Compliance
- Compatibility
- Reliability
- Serviceability
- Documentation
- Marketing
- Sales
- Packaging
- ASIC Design (Application-Specific Integrated Circuit)
- Diagnostics
- Manufacturing
- Legal
- Firmware
- Eng. Release Mgmt.

Execution Phase – EVT (Extreme Value Theory)

Purpose: Validate all H/W & S/W features exist and function

> **H/W:** Prototypes, Power-on, Functional Testing "Soft-Tooled" Enclosure, Prototype Printed Circuit Boards, System Assembled in Eng. lab, Lower CPU & Graphics speeds, Lower Clock Speeds, H/W, S/W, and Mechanical Bugs
>
> **S/W:** Feature Complete, Pre-Integration Testing Maturity

Action Initial: EMC (FCC) scans, Thermal Testing, Safety (UL, CSA) Review, Signal Integrity, OS & S/W App Testing, Mechanical Fit, Design of Experiments, Bug fixing, Alpha testing

Y/N Phase Exit Review: approval to begin next prototype build

Execution Phase – DVT (Design verification Testing)

Purpose: Validate features perform within specifications; Systems built in factory environment

> **H/W:** Electrical margin testing (four corner, HALT = [Highly Accelerated Life Testing] etc.,) Mechanical testing (shock, vibration, etc.)
>
> **S/W:** Software Quality Assurance testing of OS & Apps Maturity
>
> **Production Quality**: "Hard Tooled" Enclosure, Production quality Printed Circuit Boards, System assembly in the factory, Full speed CPU and Graphics, Full clock speeds

Action: Complete all validation and certification testing EMC, Safety, Mechanical, Reliability (MTBF demo,) Beta testing

Y/N Phase Exit: Review approval to begin Pilot Build & Mass Production (pending resolution of specific performance or quality issues)

Execution Phase - Pilot & PVT (Production Verification Testing)

Purpose: Validate Manufacturing processes assure quality goals are met; "Ownership" of the product transfers from Engineering to Operations

Action: Pilot production build (FRU's & Warehouses); Extended "burn-in" testing, Post Pack Audits; Test of order placement to shop floor Configure-To-Order processes (FRU = Field Replaceable Units)

Y/N Approval: go live with on-line ordering and begin mass production

#2. Idea Screening

This step is crucial to ensure that unsuitable ideas, for whatever reason, are rejected as soon as possible. Ideas need to be considered objectively, ideally by a group or committee.

Specific screening criteria need to be set for this stage, looking at ROI (Return On Investment,) affordability and market potential. These questions need to be considered carefully, to avoid product failure after considerable investment down the line.

#3. Concept Development & Testing

You have an idea and it's passed the screening stage. However, internal opinion isn't the most important. You need to ask the people that matter – your customers.

Using a small group of your true customer base – those that convert – the idea need to be tested to see their reaction. The idea should now be a concept, with enough in-depth information that the consumer can visualize it.

- Do they understand the concept?
- Do they want or need it?

This stage gives you a chance to develop the concept further, considering their feedback, but also to start thinking about what your Marketing message will be.

#4. Business Analysis

Once the concept has been tested and finalized, a business case needs to be put together to assess whether the new product/service will be profitable. This should include a detailed Marketing strategy, highlighting the target market, product positioning and the Marketing mix that will be used. This analysis needs to include whether there is a demand for the product, a full appraisal of the costs, competition and identification of a break-even point.

#5. Product Development

If the new product is approved, it will be passed to the technical and Marketing development stage. This is when a prototype or a limited production model will be created.

This means you can investigate exact design & specifications and any Manufacturing methods, but also gives something tangible for consumer testing, for feedback on specifics like look, feel and packaging for example.

#6. Test Marketing

Test Marketing (or market testing) is different to concept or consumer testing, in that it introduces the prototype product following the proposed Marketing plan as whole rather than individual elements.

This process is required to validate the whole concept and is used for further refinement of all elements, from product to Marketing message.

#7. Commercialization

When the concept has been developed and tested, final decisions need to be made to move the product to its launch into the market. Pricing and Marketing plans need to be finalized and the sales teams and distribution briefed, so that the product and company is ready for the final stage.

#8. Launch

A detailed launch plan is needed for this stage to run smoothly and to have maximum impact. It should include decisions surrounding when and where to launch to target your primary consumer group. Finally, in order to learn from any mistakes made, a review of the market performance is needed to access the success of the project.

Section 2: Failure Mode Analysis (FMA) & Failure Mode Effects Analysis (FMEA) as Applied to Electro-Mechanical Systems

Abstract: During the course of every system design, it is good Engineering practice to evaluate the potential failure points of each system element (See **Note:** 1); and to understand the effects of those failures on the successful operation of the system.

Although this may seem to be a mundane exercise, the benefits are certainly tangible – those benefits being increased reliability and confidence in system interconnect design to maintain signal integrity.

In order to perform this analysis, the following tools are needed:
1. Definition of "failure"
2. Types of Failures categorized
3. Some Causes of Failure
4. A suggested Approach
5. Failure Reduction Precedence Sequence

Note: 1: An element may be as small as a solder joint or as large as an entire peripheral device. To determine the element size for consideration it is necessary to understand the mission profile of the device, simply: "What must this equipment do to perform successfully" Each phase of this profile is broken down into many sub-phases, which additionally are reduced to events and sequences. All events and sequences are listed with their associated hardware and analyzed as to failures, potential failures and related failures.

1.0 FAILURE (Definitions)
The categorization of system element failure may take three forms:
- A. <u>Failure</u> (also: <u>complete failure</u>) - the act or condition of failing - fail: to be deficient or unsuccessful, to decline, weaken or cease to function
- B. <u>Potential Failure</u> - a condition which if uncorrected or unnoticed will, in some finite time frame result in a complete failure.
- C. <u>Related Failure</u> - a condition caused by an interconnected element which has failed or will fail due to a potential failure.

2.0 TYPES OF FAILURE
Each system element failure may be categorized as to its severity:
- A. <u>Catastrophic</u> - No time or means are available for corrective action (SYSTEM IS INOP)

B. <u>Critical</u> - May be counteracted by emergency action performed in a timely manner (either automatic or manual)
C. <u>Controlled</u> - Has been counteracted by appropriate design, safety devices, alarm/caution and warning devices, or special automatic/manual procedures.

3.0 CAUSES OF FAILURE

Causes of failures may be generalized or specific, but for the most part fall under one of the following instances: (It should also be noted that "failures beget failures" in a cascading effect and the root problem might not be obvious when multiple system failures occur).

A. Operator Errors - either: hardware or software
B. Design Characteristics - usually hardware
C. Procedural Deficiencies - usually software
D. Subsystem Malfunction - either: hardware or software
E. Environmental Conditions (extremes exceeded)

4.0 APPROACH

Failure Analysis Assessment consists of two activities:

A. Those techniques and analytical tools used to identify failures.
B. The process of evaluating the risks associated with the identified failures.

The first activity is accomplished most simply by the Electro-Mechanical Packaging Engineer assuming "what if" with a variety of failures and failure causes and categorizing the results.

The second activity is accomplished by comparing project goals, i.e. "deliverables" against budget constraints, work force levels, time to market, reliability goals, repair strategies, and other pertinent factors as applicable.

Also, in large systems, for a simplified method of summarizing the analytical "building blocks" that compose the failure identification process, consider the following expression:

(Failures arising from configurations) + (Failures resulting from environments/operations) = INTEGRATED FAILURES FOR SYSTEM [3]

A general format for failure analysis could consist of the following:

---

3 NASA Tech Brief, WINTER 1980 Vol. 5, No. 4. MSC-18745

A. <u>Element</u> failure identification.[4]

B. Failures arising at <u>element interfaces</u>.

C. Failures associated with software and <u>configuration unique</u> weaknesses that may contain failures.

The superimposing on these configuration design solutions, the operational, and operator interface <u>variables</u>.

There are several ways to measure risks for the purposes of a risk assessment. The simplest method consists of providing probabilities for the occurrence of a failure, or other undesired event, and arriving at a cumulative level of risk for the entire system. A second method is the widely used fault tree logic diagram with attendant probabilities. Both methods suffer the same weaknesses however (viz., risk acceptability that depends on probabilities, is (in most cases,) somewhat specious due to emerging technological advances common to R&D efforts).

5.0 FAILURE REDUCTION PRECEDENCE SEQUENCE

To eliminate or control failures, the following sequence or combination of items shall be used:

A. <u>Design for minimum failure</u>: The major goal throughout the design phase shall be to insure inherent low failure rates through the selection of appropriate design features as fail operational/fail safe combinations. Failures shall be eliminated by design where possible.

B. <u>Safety Devices</u>: Known failures which cannot be eliminated through design selection shall be reduced to an acceptable level using appropriate safety devices as of the system, subsystem, or element.

C. <u>Warning devices:</u> Where it is not possible to preclude the existence or occurrence of a known failure devices shall be employed for the timely detection of the condition and the generation of an adequate warning signal.

D. <u>Special Procedures:</u> Where it is not possible to reduce the magnitude of an existing or potential failure through design, or

---

4 For each event the analytical process explores those factors that could result in an event occurring prematurely, failing to occur, occurring out of sequence and several other postulated off-nominal occurrences. Each of these is then scrutinized by a team of specialists to discern any elements of failure associated with the event.

the use of safety and warning devices, special procedures shall be developed to counter failure conditions for enhanced reliability.

Product failure ranges from failure to sell the product to "fracture" of the product, in the worst cases leading to personal injury, the province of forensic Engineering.

Although product failure causes both marketers and consumers to incur substantial damage and losses, failures are often very difficult for marketers to control. The argument given is that the "cost trade-off" analysis failed to provide a reasonable estimate of what a failure would actually cost and the potential damage to the user in terms of personal injury or death. (If your cell-phone "dies" – you are sad, if your brakes fail – you are "dead"!)

After a product failure experience, consumers responded with the least favorable evaluation for brand-caused failure, a more favorable evaluation for natural disaster-caused failure, and the most favorable evaluation for consumer-caused failure. However, outcome severity moderated the effects: When the failure resulted in a severe outcome, positive brand evaluation deteriorated in the case of consumer-caused failure only. In addition, brand-blame attribution mediated these relationships. [5]

---

[5] *Effects of Product Failure Severity and Locus of Causality on Consumers' Brand Evaluation:* Song, Sujin; Sheinin, Dan A.; Yoon, Sukki: Social Behavior and Personality: an international journal, Volume 44, Number 7, 2016, pp. 1209-1221(13) Scientific Journal Publishers

Section 3: Evaluation Procedure for Cabinet/Enclosure Design

Abstract: Electromagnetic interference (EMI,) also called radio-frequency interference (RFI) when in the radio frequency spectrum, is a disturbance generated by an external source that affects an electrical circuit by electromagnetic induction, electrostatic coupling, or conduction. The disturbance may degrade the performance of the circuit or even stop it from functioning. In the case of a data path, these effects can range from an increase in error rate to a total loss of the data. Both man-made and natural sources generate changing electrical currents and voltages that can cause EMI: automobile ignition systems, cell phones, thunderstorms, the Sun, and the Northern Lights. EMI frequently affects AM radios. It can also affect cell phones, FM radios, and televisions. EMI can (also) be used intentionally – for radio jamming, as in electronic warfare. Since the earliest days of radio communications, the negative effects of interference from both intentional and unintentional transmissions have been felt and the need to manage the radio frequency spectrum became apparent.

In 1933, a meeting of the International Electro-Technical Commission (IEC) in Paris recommended the International Special Committee on Radio Interference (CISPR) be set up to deal with the emerging problem of EMI. CISPR subsequently produced technical publications covering measurement and test techniques and recommended emission and immunity limits. These have evolved over the decades and form the basis of much of the world's EMC regulations today.

In 1979, legal limits were imposed on electromagnetic emissions from all digital equipment by the FCC in the USA in response to the increased number of digital systems that were interfering with wired and radio communications. Test methods and limits were based on CISPR publications, although similar limits were already enforced in parts of Europe.

In the mid-1980s, the European Union member states adopted a number of "new approach" directives with the intention of standardizing technical requirements for products so that they do not become a barrier to trade within the EC. One of these was the EMC Directive (89/336/EC) and it applies to all equipment placed on the market or taken into service. Its scope covers all apparatus "liable to cause electromagnetic disturbance or the performance of which is liable to be affected by such disturbance."

This was the first time there was a legal requirement on immunity, as well as emissions on apparatus intended for the general population. And although there may be additional costs involved for some products to give them a known level of immunity, it increases their perceived quality as they are able to co-exist with apparatus in the active EM environment of modern times and with fewer problems.

Many countries now have similar requirements for products to meet some level of Electromagnetic Compatibility (EMC) regulation.

One strategy that makes it possible for Data Processing Equipment to meet the current worldwide RFI/EMI requirements is known as: CONTAINMENT. In that approach, any radiated emissions are contained by the nature of the enclosure structure, which assumes an electrical component status by adsorbing and grounding spurious signals and preventing their escape to the environment. All exterior surfaces of the enclosure are electrically bonded to achieve a complete shield around the emission producing components.

Inherent with the Containment approach are the problems associated with joining large numbers of tolerance-dependent pieces in a tolerance-dependent Manufacturing environment. This method will result in some improper alignment of enclosure parts, which in itself, will cause increases in opening/slot sizes and or loss of seal efficiency.

This Section explains methods available to evaluate the success of enclosure design and some measurement criteria to be used in determining whether a newly designed enclosure will provide the necessary lifetime quality production that is required for products sold into the marketplace where RFI/EMI is a concern.

CONTAINMENT: Using a "CONTAINMENT" approach assumes the radiated emissions are contained by the properties of the enclosure structure which assumes the status of an electrical component by adsorbing and grounding spurious signals and thus, preventing their escape to the environment. This is accomplished by screening or gasketing all seams or orfices which would allow interfering frequencies to escape. Various methods are available to determine radiated frequencies versus the slot/opening size(s,) but this Section will not deal with those formulae. The assumption is made that the shielding design is sufficient to allow proper attenuation of the equipment generated frequencies. (As stated before, shielding effectiveness is proportional to the size of the openings.)

From the above, it may be seen that any increase of opening size, or loss of seal efficiency, may cause excessive RFI/EMI emissions, and failure to meet the emissions criteria for that product.

Inherent with the Containment approach are problems associated with joining large numbers of tolerance dependent metal pieces in a tolerance dependent Manufacturing environment. This method may result in improper alignment of enclosure parts, which will cause dimensional changes in openings or slots and or loss of seal efficiency. The term, which may be used to express this, is: "Joint Unevenness" – a Discontinuity of Planar, Parallel, Surfaces. Several causes of joint unevenness are:

- Broken Parts
- Thermal Effects On Thin Metal Pieces
- Bent (Incorrectly) Pieces
- Shipping Damage
- "Loss" Of Assemblies (It "Fell" Out/Off)
- Manufacturing Tolerances
- Assembly Tolerances/Errors
- Lack of Proper Closure Pressure Vs. Resiliency
- Material Relaxation - Beyond Yield Limit

There are time and maintenance cycle dependent problems associated with the Containment approach, which are related to the electrical contact portion of the enclosure seal(s). These effects may be categorized as follows:

Contact Resistance Increases

A. Corrosion Effects
B. Lack of Pressure Closure
C. Maintenance Cycling Effects On Shielding

These effects are often subtle in nature and do not lend themselves to visual checks. These effects may require empirical test data to evaluate.

Fortunately, all the factors mentioned are measurable in some form, the errors/tolerances are controllable, and most of the design problems are correctable. The issue is that in any given enclosure design, the potential for loss of seal efficiency or "Joint Unevenness" is unquantified/unknown until there is sufficient test data gathered to successfully evaluate the design. There are no guarantees that one method is superior to the other, (or will satisfy all applications.) A design could be excellent in test, but a failure under field use.

Many types of rigorous test programs may be applied to cabinet/enclosure design to obtain some level of confidence in the initial design, and predict failure modes for correction. It is unwise to say that such rigorous testing would be the norm, but often the best procedure is some testing followed by an analysis of selected populations in the Field installed base. Testing processes, which may be germane to this data development and some test plans and philosophies, are presented in the following discussion.

One of the prime considerations in developing a comprehensive approach to mechanical reliability analysis must be the effective use of a broad spectrum of reliability information.

The following flowchart provides an example of the types of testing/examination that may be specified to assure the cost/quality goals of the product design are met. It is not intended to be all-encompassing or all-inclusive.

Lack of testing/analysis opens the door to "Consumer-use surprises," realized when the products, as used, fail to meet expectations or create emissions above and beyond those allowed by statute.

FIGURE 1: Flowchart Example of Rigorous, Controlled, Testing

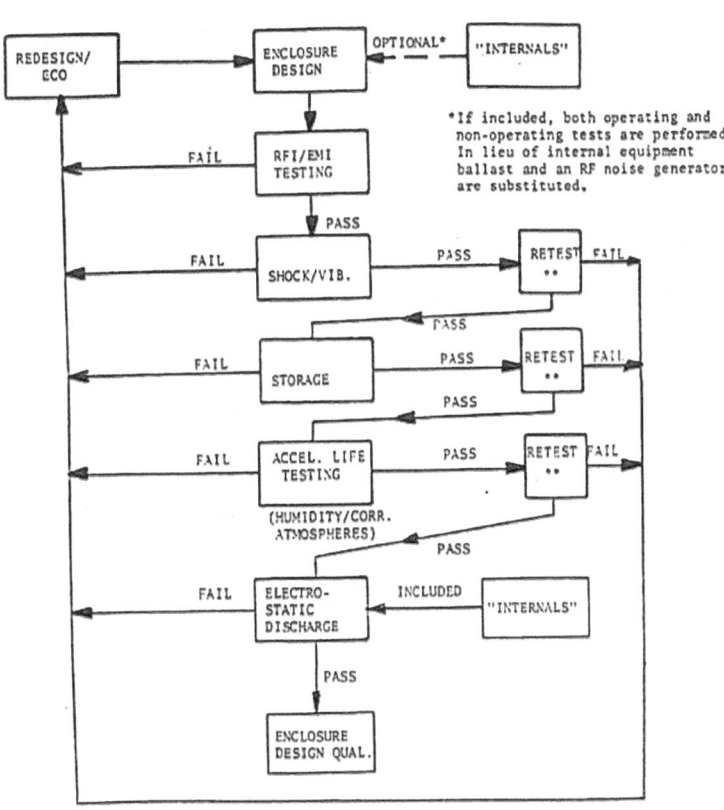

**Emission/Susceptability

Significant points that must be considered for the mechanical reliability analysis are: [6]

   A. Specifications of mechanical parts and elements to be considered in the reliability study.

   B. Determination of failure modes.

   C. Determination of governing failure mechanisms.

   D. Analytical description of failure mechanisms, if feasible.

   E. Use of statistical information on failure modes or mechanisms, if such information is available.

   F. Determination of stresses on each part.

---

6 IITRI project E6019, "METHODS FOR PREDICTION OF ELECTRO-MECHANICAL SYSTEM RELIABILITY" by Thomas L. Bush, Anthony P. Meyers, & Darwin F. Simonaitis. Published in the Semi-annual report for the U.S. Army Electronics Laboratory@ Ft. Monmouth N.J., 15 May 1964 to 14 November 1964; pages 46 and 47.

G. Estimation of reliability characteristics from part stress-failure mechanism relationships.

H. Qualitative evaluation of part reliability where necessary.

I. Determination of steps to be taken to improve reliability where necessary.

J. Determination of design, application, and operational factors, which can affect performance and reliability.

K. Determination of mechanical system reliability descriptors and interpretations.

L. Calculation of total system reliability parameters.

INSTALLATION TESTING – TEST 1:(See Note on the following page)
Installing A Manufactured Compliant Product

These points should be demonstrated: (As applicable – based on system size and design.)

a. A large enclosure should be installable by one person using common hand tools.

b. During installation/movement, the enclosure, shall not tip or exhibit physical instability.

c. Access to cable connector bulkheads.

d. Access to air filters.

e. The joining of two cabinets/enclosures shall not require more than XX minutes.

f. Demonstrated Tip Angle to be:) < 15 degrees from normal.

TEST 2: (See Note on the following page)
Mating of a "Compliant" Component with a "Non-Compliant" Enclosure(s,) Or Attempt to Upgrade Equipment in The Field.

Using selected kits perform points a, b, c and e, per TEST 1 with available documentation.

Keep in mind, that even an "individual user" can "screw-up" a product through mis-use, abuse or sheer ignorance of operating procedures – making their "problem" your "problem."

SHIPPING & HANDLING: TESTING

The Shipping Material Demonstration should consist of a dummy loaded enclosure packed for shipment. The intent is to watch the packaged component(s) as moved around a simulated site, observing the handling characteristics, uncrating the enclosure/option, and removing the enclosure/option from the shipping skid to make it ready for installation. All this is done with the documentation as provided, and with no assistance offered.

All evaluations are based on the following criteria:
  a. Person Power (weight lifted/person) to perform the tasks.
  b. Skill-level of the person(s)
  c. Safety aspects
  d. Time to perform tasks (hours/minutes)
  e. Relative Ease to perform
  f. Quality of the final assembly
  g. Readability of the documentation provided and its applicability to the actual task.

It is necessary to balance rigid test requirements with a general understanding that unless processes are somewhat guided and controlled, there will be a decreased confidence level in the basic enclosure as a viable containment for RFI/EMI emissions.

The possible risk associated with any containment failure is, if inadequate determinations of design quality are made, and a resultant RFI/EMI "complaint" is forwarded to a Corporation, then some enclosures may have to be retrofitted/modified in the field at an expense to the Corporation. The testing done to recertify the modifications may be just as costly as those tests done during product development. There is, admittedly, no metric available, which would allow ready evaluation of either approach. The cost for the field retrofits would be borne by the manufacturer and could be extended, by market demand, customer request, or edict, to other products.

Once a decision is made to use a "containment strategy" to meet emission requirements, it is also safe to assume that enclosures will assume a "component functionality" (Electrical,) which they never had before. This newfound "functionality" would have a predictable reliability associated with it, and failure rates assigned.

Note: For "Consumer-Use" products that require no support for installation sufficient testing should be developed and documented to assure that the product is "robust" enough in design to withstand unforeseen conditions and user "abuse."

The application of the standard "bathtub curve" shows approximately what failures modes could be encountered at various points in the product life.

FIGURE 2. Application of "Bathtub Curve" (Depicting Component Failure Rate as a Function of Time)

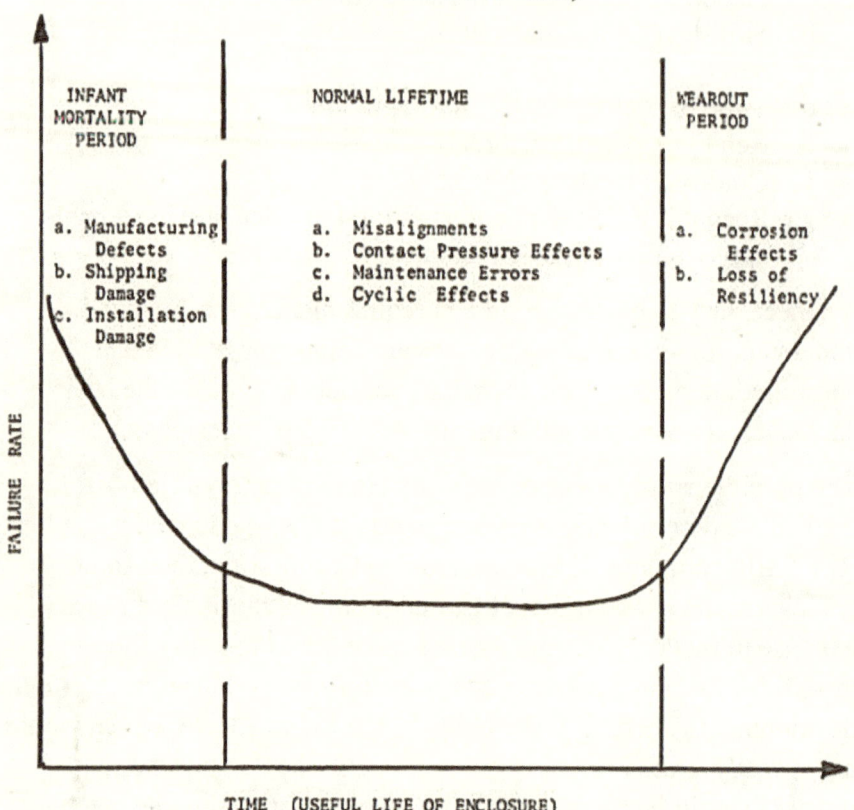

For purposes of discussion, a failure rate was assumed based on a product life of ten years @ 100% duty cycle, and constant failure. The plot is projected and modified to show the resultant effects of the various failure modes. This demonstration is for representation of the model only and is not based on actual data.

> **Note:** Now, it might be appropriate to consider that in a "perfect world" design, there would be no need for the product enclosure to act as the last line of defense against RFI/EMI, Electro-Static...etc., problems. We live in an "imperfect world," – it's up to your skill as an Engineer and Product Designer to save the world from those who cannot "silence" their circuit-noise.

FIGURE 3: Constant Failure Rate Prediction with Modifiers

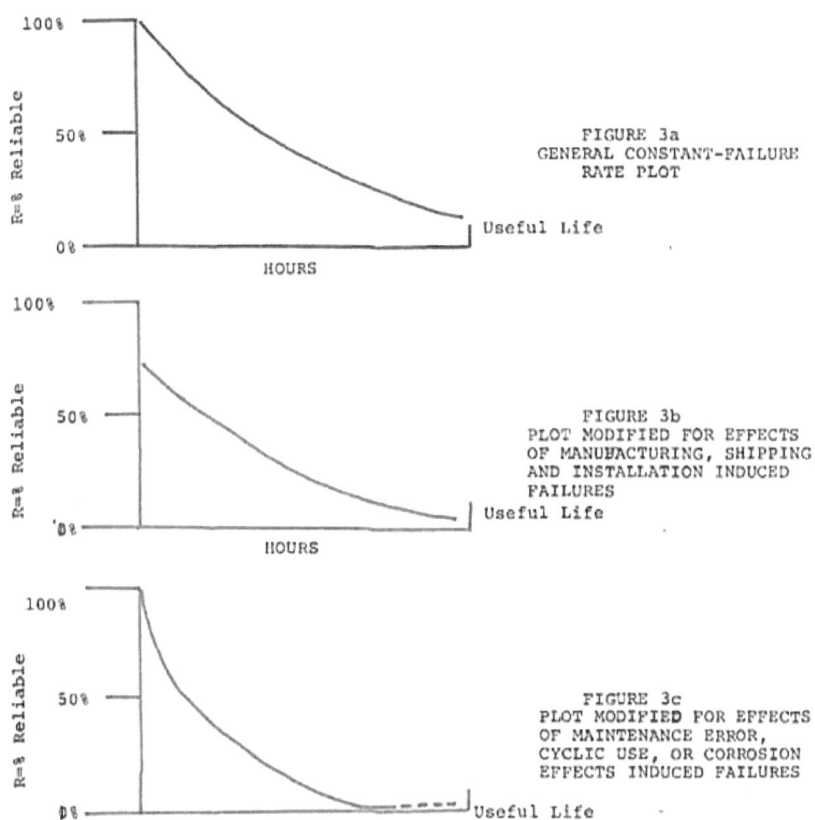

Although the "availability" of an Electro-Mechanical system may not be (directly) affected by a failure of the enclosure components to contain RFI/EMI emissions, there will be those cases where the failure of enclosure designs will result in either signal integrity problems (manifested as increased soft errors,) or in an actual non-compliance issue; necessitating down-time for measurement, analysis, and resolution of the problem.

Such a situation is not readily quantifiable, but it is a consideration in design analysis since it will tend to decrease availability as complaints or soft-errors increase. This type of problem has a complex solution. Some of the problems may be rectified quite simply by correcting faults found in installation or use, while fixes, which fall into the design-remedial case, may result in substantial downtime for problem sites.

What becomes a critical factor, throughout this exercise, is the population of enclosures to be addressed. Sales of new electronics are driving increases in the use, storage, and end-of-life management of electronics. An estimated 438 million electronic products were sold in 2009, which represents a doubling of product sales from 1997-driven by a nine-fold increase in mobile device sales.[7] This estimate will help in the following discussion of the premise that, "Electro-Mechanical Systems never fail-they last a hundred years."

The discussion does not concern itself with the structural integrity of the enclosure. The actual points of consideration are those electro-mechanical interfacing pieces, which provide for the electrical continuity of the enclosure and thus give the Containment Strategy reality. In that sense, the enclosure could very well fail over a given time period.

A determination of a theoretical failure rate was made based on one enclosure @ 100% duty cycle, which will fail one time in one hundred years equals 0.0000011 failures per hour (or 1.141 failures in one million hours). From our (estimated) population of enclosures: > 438,000,000 units, we multiply this and that answer multiplied by the hours at 100% duty cycle during any five-year sample. This yields a total of: > 21,100,000 "failed" units

There are also several assumptions, which are made with this model. First, the product quality is assumed to be 100% from Manufacturing. Second, the product quality is not effected by installation techniques, or installer skill level. Third, there are no maintenance effects or damage caused by maintenance considered. Field data could be used to modify these assumptions to yield a more accurate result.

In order to evaluate these values and place some credence on the model, it would be necessary to monitor sample sites under controlled

---

7 *America Goes Green: An Encyclopedia of Eco-Friendly Culture in the United States*; Volume I: Thematic Entries, Kim Kennedy White, ABC – CLIO, LLC., 2003, Page 675

conditions and to report problems encountered with enclosures as they occur: during installation, during maintenance or during use.

It will be necessary to report any suspected or confirmed RFI/EMI emission problems so that a database may be formulated to ascertain the scope of the issues. If these problems are of insignificant levels relative to the population, then a high degree of confidence could be placed on the techniques used to design and assemble products. If a large number of occurrences are noted, it may be necessary to modify some aspects of the containment strategy and to validate changes and future designs with test data.

In summation, the typical failure modes expected to be seen when using a Containment approach to RFI/EMI emission regulation compliance have been listed and techniques for evaluating their reliability have been discussed. The use of this strategy for compliance causes the enclosure to assume a new character, which should be of consideration in future designs.

> **REPEAT:** Now, it might be appropriate to consider that in a "perfect world" design, there would be no need for the product enclosure to act as the last line of defense against RFI/EMI, Electro-Static...etc., problems. We live in an "imperfect world," – it's up to your skill as an Engineer and Product Designer to save the world from those who cannot "silence" their circuit-noise.

Section 4: Determining Life-Cycle Cost

Abstract: Product failure is deceptively difficult to understand. It depends not just on how customers use a product but on the intrinsic properties of each part – what it's made of and how those materials respond to wildly varying conditions. Estimating a product's lifespan is an art that even the most sophisticated manufacturers still struggle with. And it's getting harder. The consumer expects devices to continuously be getting smaller, lighter, more powerful, and more efficient.

This thinking has seeped into our expectations about many product categories: Cars must get better gas mileage. Bicycles must get lighter. (Washing machines need to get clothes cleaner with less water.) Almost every industry is expected to make major advances every year. To do this they are constantly reaching for new materials and design techniques. All this is great for innovation, but it's terrible for reliability.

Change is difficult, change is disturbing, and change brings uncertainty. Change can create failures, but it also can create success. Understanding when and why things fail is critical to a company's well-being.

Since 1980s, the reliability of electronic components (in general) has improved two to four orders of magnitude. Parts were often specified in failures per million hours of operation [the term used is λ (lambda)]. Today, parts are specified in failures per billion hours of operation, which are referred to as FITs (Failures in Time). If parts were the main contributor to failures, then, with the vastly improved complexity of new devices, they would be failing constantly.

We can all attest that they are not. Televisions, radios, and automobiles all have more parts and last longer. This is due to the inherent design and the Manufacturing processes, not the parts count. What is needed to improve the reliability of a manufactured assembly is to improve the design and the Manufacturing process. Much work has been done over the past few decades to improve the quality and reliability of components. This effort has, for the most part, been very successful. In fact, the measurement used to describe the quality of components has been changed three orders of magnitude as well (from λ to FIT).

The following truncated bibliography provides excellent resources for further research and application:

MILITARY HANDBOOK: Reliability Prediction of Electronic Equipment	
MIL-HDBK-217F	2 December 1991
Superseding: MIL-HDBK-217E, Notice 1	12 January 1990
MILITARY HANDBOOK: Electronic Reliability Design Handbook	
MIL-HDBK-338B	1 October 1998
Superseding: MIL-HDBK-338A	12 October 1988
All documents are available "FREE" in PDF Format from the following website:	
http://www.weibull.com/knowledge/milhdbk.htm	10 October 2016

You've completed your Failure Mode Analysis and you know the items that are most critical to functionality and successful operation. You (also) have an understanding of what the product "should do" in the hands of the customer.

During a design review, there is a "heated discussion" concerning component cost, (especially for one of your identified critical components.)

A suggestion is made to: "Go with the cheaper costing item, and we'll handle any failures by way of a customer service call/repair, or (worse yet,) a return swap."

Aside from the obvious degradation of design quality, customer satisfaction, and perception of the product in the marketplace there is a legitimate cost associated with that approach.

The following discussion is an explanation of how you can quantify the life-cycle cost of a component based upon the expected number of failures in each year. (Be sure to include the "population effects" as more of the product is shipped to the field and, perhaps revisit the "aging effects" that lower quality components will have on overall product reliability. It is true, "Nothing lasts forever," but is also true that the customer expects to "Get their money's worth" out of the product.

DERIVATION: For the parameters relevant to this model; the life-cycle cost (LCC) of a component is equal to:

$$LCC = C + \frac{S_1 P_1}{(D)_1} + \frac{S_2 P_2}{(D)_2} + \cdots\cdots\cdots\cdots \frac{S_i P_i}{(D)_i}$$

Where:

- C = Component cost
- P_i = expected number of failures in year i
- S_i = cost of a service call in year i
- D = discount factor

If we then:

1. Remember that: $P_i = P_i + 1$, due to the exponential distribution of failure rates in almost all components
2. Assume a $300 constant cost per service call

> (You must insert your own costs). Note: If there are "NO" service call-costs, you should insert the "Cost to Return" a product for repair.

3. Assume a 20% discount factor over a five-year life (You may insert your own discount factor); the equation given above becomes:

$$LCC = (300) \left[\frac{(\lambda)(8760)}{10^6}\right] \left[\frac{1}{1.2^1} + \frac{1}{1.2^2} + \cdots \frac{1}{1.2^5}\right] + C$$

Where the term: $\left[\frac{(\lambda)(8760)}{10^6}\right]$ converts the failure rate / million hours (λ) into the expected number of failures in one year (8760 hours).

The above immediately reduces to:

$LCC = (2.628)\,(\lambda)$

Which then reduces to:

$LCC = (2.628)\,(\lambda)\,(2.99) + C$ OR: $LCC = 7.86\,(\lambda) + C$

Section 5: Errors

Abstract: Error exists. Everywhere. It cannot be avoided. Can you see this error? Can you identify the types? Do you have a strategy for dealing with each type? Can you figure out which source contributes the most to the error in the final result? Do you attempt to wrestle with these questions?

Probable Error: Manufacturers of anything know the probable error. Otherwise they could not ship it. Often the error is documented with the product.

Systematic Error: Any measurement device can be used inappropriately. A math error, an error of when, where or what can lead to systematic errors. (The fog associated with frustration is often systematic error.) A working system can appear broken because it is hidden behind a combination of fogged thinking and inconsistent use of measuring equipment. Someone else, with hardly any understanding, can ask a single question that causes the systematic error to disappear.

Systematic errors can disappear suddenly without apparent explanation. This is not time to rejoice, but for more puzzlement. What changed? Something changed. (It is now working.) Systematic error can be associated with an operator of the equipment. Unconscious feelings, expectations, and habits can cause a person to create systematic errors. Battle this by starting over from scratch. Recreate the problem. Now repeat what you think happened to see if it starts working again.

Most equipment has quirks of detail that need to be appreciated. For example, there is a mirror behind the needle of most meter movements. The designers' assumption is that a user will move their head so that the needle image in the mirror is directly underneath the needle.

Random Error: Random errors cannot be fixed. A better word is "uncertainty." Not only has the quantum uncertainty that underlies nature been experimentally verified, it has been proven both mathematically and experimentally that we cannot know more.

The problem with uncertainty is that it grows as more complex systems interact. The uncertainty, or error, can overwhelm the predictability of physics and chemistry.

Random error is assumed because it is a first best guess, a maximum worst case possibility, a starting point, an easy place from which to improve error analysis.

Repeated Measurement Error Reduction: Repeating the same mistake (or systematic error) over and over again is silly. Suppose someone is measuring length but is moving their body and head into a different random position each time (very unlikely to be random). Repeatedly measuring length over and over again and using statistics may reduce the error. But this is highly unlikely.

When to Assume a Random Error: Assume random error when the fog around failure is unknown. For instance, the experiment yields different results every time it is repeated. Perhaps one or perhaps 10 systematic mistakes are being made. The random assumption is probably not true, but assuming random error as a first approximation can determine if a second experiment was better or worse.
The objective of random error assumptions is not to discover the source of systematic error, but to find a starting point to begin searching for systematic errors. The goal is to build confidence and figure out the next step and by estimating error.

Leave it to NIST (National Institute of Standards and Technology) to identify the truly random error and more accurate decimal places. Use Random Error at the beginning of the project because of human frailty and to quickly compute maximum error. Expect scientists to use Random Error to account for the uncertainty of nature weirdness at the end of the project.

Instrument Error: Most instruments are to be read with a plus and minus reading. They are not to be read over and over again as if the error is random.

Tolerance: Tolerance is space built into the design between parts. There are standards for tolerance specification and assumptions if none are mentioned. They are used by Engineers to communicate technical detail to those machining or building the parts.

Allowance: Allowance is another example of how an Engineer communicates to those making parts. Allowance describes the "extra" to be left in for a specific purpose either during assembly or later during the part making process.

Calibration: The best measuring tools will come with a certificate, a piece of paper certifying Calibration. Some will come with a sticker, some a notebook. Some never need their calibration checked in the future. Some need to be checked every year or every 6 months like an elevator. Every measurement device has its own standards that describe how to calibrate it.

Checking calibration can be a process that is done by certified technicians in-house or may need to be shipped somewhere. The logistics of calibration can double the equipment cost and significantly delay a project if not considered beforehand.

Why all this concern over "Errors" Why should an Electro-Mechanical Packaging Engineer worry about the effects of test equipment errors Simply put, **Murphy's Law** – an adage or epigram that is typically stated as: *Anything that can go wrong, will go wrong.*

During the creation of an Engineering Product Specification considerable care must be given to understanding the effect of assigning testing values that are unrealistic or unattainable.
- Know what you are testing.
- Understand the testing procedure.
- Specify a reasonable, reproducible result.
- Unless you are working for NASA or the DoD – there ARE "cost constraints.

Give yourself, and your design some "slack" – allow the design parameters to encompass the requirements – don't "strangle the requirements."

Live with errors you can and design-out errors you can't "stand".

Section 6: Sample Product Requirements

Abstract: In establishing Product requirements, the first step is to define the "needs" for the requirement. The needs are based on a want or desire. Usually, an individual or organization identifies a need for an item or function early in the design process. (This may in the form of a customer specification document, a set of "understood" Marketing needs, or a general understanding of the end-use of the product in its market.) After a need is defined, feasibility studies should be conducted to evaluate various technical approaches that can be taken. The Product operational requirements should also be defined. This includes the definition of Product operating characteristics, maintenance support concept for the Product, and the identification of specific design criteria. In particular, the Product operational requirements should include the following elements.

- Mission definition: Identification of the primary operating mission of the Product in addition to alternative and secondary missions.
- Performance and physical parameters: Definition of the operating characteristics or functions of the Product.
- Use requirements: Anticipated use of the Product.
- Operational Distribution: Identification of transportation and mobility requirements. Includes quantity of equipment, personnel, etc. and geographical location.
- Operational life cycle: Anticipated time that the Product will be in operational use.
- Effectiveness factors: Numbers specified for Product requirements.
  - Includes Cost/Product effectiveness
  - Mean time between maintenance (MTBM)
  - Failure rate (MTTF)
  - Maintenance downtime
  - etc.
- Environment - Definition of the environment in which the Product is expected to operate.

Basically, the Product operational requirements define how the Product will be used in the field by the customer.

Usually, in defining Product requirements, the tendency is to cover areas that are related to performance as opposed to areas that are related to support. However, this means that emphasis is only placed on part of the Product and not the whole Product. It is essential to take into consideration the entire Product when defining Product requirements. (The Product Maintenance concept basically describes the overall support environment that the product is supposed to exist in.)

After the Product operational requirements and Product Maintenance concept are defined, a preliminary Product analysis is performed to determine which approach for Product development should be adopted. The following process is usually applied.

1. Define the problem - The first step always begins with clarifying the objectives, defining the concerned issues, and limiting the design problem so that it can be effectively studied.
2. Identify feasible alternatives - All the alternatives should be considered to make sure that the best approach is chosen. (Based upon specified values).
3. Select the evaluation criteria - The criteria for the evaluation process can vary considerably, so the appropriate ones must be chosen.
4. Generate Specification statements/data - The requirements data should be specified.

The results will be defined as either technical requirements included in the specifications, or management requirements included in a program management plan.

Note: A Product Engineering Management Plan contains three sections.

1. The technical program planning and control part describes the program tasks that have to be planned and developed to meet Product Engineering objectives such as work breakdown structure, organization, risk management, etc.
2. The Product Engineering process part describes how the Product Engineering process applies to program requirements.
3. Finally, the Engineering specialty integration part describes the major Product-level requirements in the Engineering specialty areas such as reliability, maintainability, quality assurance, etc.

The Product Specification includes information from the operational requirements, maintenance concept, and feasibility analysis.

It is essential to establish the correct requirements and specifications early in the development process to prevent errors later on in the Product Development cycle. However, the Electro-Mechanical Packaging Engineer still needs to realize the reality of requirements changing over time.

There are three reasons why it is difficult to make good requirements.

- First, the requirements for an embedded Product are even more complex than for a "stand-alone" Product because an embedded Product must interact with the outside world and not just other components.
- Second, establishing correct requirements requires people with both technical and communication skills.
- Third, the complexity of establishing correct requirements makes it more of an art than an Engineering skill. Often people have the attitude of "I'll know it when I see it," thus making it difficult to establish the requirements early.

Note: The examples are that follow are strictly for illustrative purposes only and are not intended to be representative of any actual product requirements.

Your (actual) product requirements will vary, or by co-incidence, match some of the illustrations. The statements made in these examples are not to be interpreted as design guidance or a guarantee of expected Product performance. You should be aware of the risks of adopting specifications without carefully considering the input from other design-team members, especially before making any final decisions.

You cannot assume that any recommendations made in any part of this document will meet your Product requirements, except by co-incidence and the contents of this document are provided for information/instructional purposes only.

Remember: When you write these documents, you must be thorough, but also concise. For example:

- Always use the approved Corporate "Word-Processing" format, including any special templates, (if available.)
- Always put a date stamp at the top of the pages, (the header) showing when the file was most recently updated.
- Maintain a "revision" page.
- Keep copies of "Old Revisions" in a Project Folder

The first document you should create describes the data that shall be included in a Product Engineering Specification to provide a source for consistent and accurate information about (all) [*Insert Your Company Here*] products. The document will permit Engineers/Test Personnel the ability to collate all relevant parameters about a product in one place (the Engineering Specification,) and make this information available to all [*Insert Your Company Here*] personnel who deal with customers. Usually 8 ½ x 11 paper, the document contains:

A Title Page   (ON ITS OWN PAGE)

A Revision History/Status   (ON ITS OWN PAGE)

A Table of Contents   (ON ITS OWN PAGE)

1. INTRODUCTION
   - 1.1. PURPOSE
   - 1.2. SCOPE
   - 1.3. RESPONSIBILITIES
   - 1.4. REFERENCED STANDARDS
   - 1.5. CONFORMANCE
   - 1.6. GENERAL METHODS
     - 1.6.1. Test Units
     - 1.6.2. Measurement Error Considerations
     - 1.6.3. Preliminary Data
     - 1.6.4. Recording of Data
     - 1.6.5. Media
     - 1.6.6. AC Output Current Devices
     - 1.6.7. Non Applicable Parameters
     - 1.6.8. Additional Parameters
     - 1.6.9. Measurement Units

2. PHYSICAL SPECIFICATIONS

3. ENVIRONMENTAL SPECIFICATIONS

4. AC INPUT POWER SPECIFICATIONS

5. DC INPUT POWER SPECIFICATIONS

6. ELECTRO MAGNETIC COMPATIBILITY (EMC) SPECIFICATIONS

7. REGULATORY AGENCY COMPLIANCE

8. Appendices – Specification Tables

Revision History/Status (A simple table will suffice) – See Below:

Date	Version	Description of Change
10/14/2016	0	Initial Release
10/15/2016	1	Added regulatory Agencies to Section 7.0
10/16/2016	2	Added Notes and Appendices
...	...	...

1. INTRODUCTION   (START A NEW PAGE)

The [*Insert Your Company Here*] provides its customers with information about products that is sometimes conflicting or even erroneous. To reach the customer, this information must flow from Engineering to Configuration Groups, Field Service, Marketing, Educational Services Development and Publishing, and Customer Services. From the Engineer to the customer, there must be a clearly defined vehicle by which clear, comprehensive information can be transferred.

This Document describes the data that shall be included in a Product Engineering Specification to provide a source for consistent and accurate information about products. It permits Engineering to collate all relevant parameters about a product in one place (the Engineering Specification,) and make this information available to all [*Insert Your Company Here*] personnel who deal with customers.

1.1 PURPOSE

This section defines the minimum electrical, physical, and environmental parameters that must be recorded, during testing, in a product Engineering Specification. This information is needed by [*Insert Your Company Here*] personnel to accurately perform such tasks as preparing computer sites, writing sales literature, and creating hardware installation and operation manuals. Once the information is compiled as a Product Engineering Specification, it shall be accessible to all [*Insert Your Company Here*] personnel.

1.2 SCOPE

This section outlines the parameters needed to properly specify a product's electrical, environmental, and physical characteristics. These parameters are grouped into generic categories. Some categories apply to a particular product while others do not. (See subhead 1.5 to determine which categories are applicable.) These parameters represent minimum requirements. The appendix provides tables to record parameters as they are defined in subheads 2 through 6. Cables, media, software, and packaged systems are not addressed by this section. However, all products contained in a packaged system shall be specified according to this standard.

> **Note:** Parameters covered here do not include product performance specifications.

1.3 RESPONSIBILITIES

1.3.1 Equipment Design Engineers

Equipment design Engineers, at the option level, are responsible for using this standard to define product Engineering Specifications. They are responsible for completing and updating all of the tables contained herein as they apply to a particular product. Further, it is the responsibility of the option design Engineer to verify or obtain, by testing or measurement, all of the appropriate parameters in the product Engineering Specification.

1.3.2 Users of Product Engineering Specifications

Users of product Engineering Specifications (Marketing, Educational Services Development and Publishing, Field Service, facility Engineers, etc.) are responsible for using current Product Engineering Specifications as the source for electrical, physical, and environmental parameters for products

1.4 REFERENCED STANDARDS   (START A NEW PAGE)

Usually contains a list of [*Insert Your Company Here*] standards.

1.5 CONFORMANCE   (START A NEW PAGE)

Any product that is listed in the Option Module List must have all applicable specification tables completed. Exceptions are noted in subhead 1.2. The following table can be used to determine which tables must be completed for your product.

SUBHEAD	PRODUCT TYPE
2 Physical	Any product, equipment, hardware, module, or option.
3 Environmental	Any product (including media used by the product,) that has temperature, humidity, vibration, or other environmental constraints (i.e., any product tested)
4 AC Input Power	Any product that has an AC line cord or connects to utility power
5 DC Input Power	Any product that uses an external DC power source derived from power supplies or batteries.
6 EMC	Any product that is tested
7 Regulatory	Any product that has been designed or tested for compliance to any regulatory standard.

1.6 GENERAL METHODS

1.6.1 Test Units: A minimum of five units is recommended for obtaining final test data. For certain tests, such as shock and vibration, altitude, EMI, etc., final data can be obtained by testing only one unit. This section contains a variety or tables for record ing minimum, typical, and maximum values. Minimum values in the table represent the minimum data obtained from the test units. Typical values in the table should be obtained by averaging the data from the basic test units. Maximum values in the table should be maximum data obtained from any test unit at maximum power supply rated load.

1.6.2 Measurement Error Considerations: Most of the specifications addressed in this section must be obtained or verified by measurement. A recommended over all measurement accuracy is given for each parameter. The actual measurement accuracy for a test depends upon the instruments used. The actual measurement error (assumed to equal the measurement accuracy) must be accounted for in obtaining or verifying specifications. Actual measurement error should be applied to make the data recorded in the Product Engineering Specification conservative.

1.6.2.1 Specifications Obtained by Measurement – Data obtained by measurement (dimensions, air-flow, heat dissipation, acoustics, AC and DC currents, AC and DC power, etc.) must be adjusted by the actual measurement error to make the data conservative.

For example, if inrush current is measured with ± 6% accuracy, readings must be increased by 6% to compensate for this error, thereby making the data conservative.

If a product or skid dimension is measured with ±2% accuracy, the dimension should be increased by 2%. In another example, if input power is measured with ±4% accuracy, the data must be increased by 4% to compensate for the error.

And finally, if air-flow or output watts are measured with ±4% accuracy, the data must be decreased by 4% to make the data conservative.

1.6.2.2 Specifications Verified by Measurement – Those specifications that are verified by measurement (temperature, humidity, voltage and frequency range, mechanical shock and vibration, output wattage, etc.) should have measurements with margins equal to or greater than the actual measurement error.

For example, if relative humidity of 20% to 60% is verified with ±3% measurement accuracy, the instrument should indicate 17% to 83% relative humidity.

In another example, if a voltage range of 90 to 128 volts is verified with a ±2% measurement accuracy, the instrument should indicate 88 to 131 volts.

Finally, if a frequency range of 47 to 53 hertz is verified with ±1/2 hertz measurement accuracy, the instrument should indicate 46.5 to 53.5 hertz.

> **Note:** Poor accuracy instruments can require the product to be verified with wide margins. When a product fails because of this effect, an instrument with better accuracy may be used to decrease the test condition margins.

1.6.3 Preliminary Data: Estimated or calculated data (no measurements taken) can be used for preliminary specifications only. Such preliminary specification tables must have a "PRELIMINARY" heading on each page.

1.6.4 Recording of Data: Preliminary specification tables should be included in the Product's Engineering Specification, until replaced by Final Specification tab les. The Product's Engineering Specification must be under revision control.

1.6.5 Media: Any product that has sub-assemblies (including media such as paper, discs, and CD Drives) must include the most restrictive limits or requirements in the specification for the product.

Note: If extended range media or special media give a significantly better performance or range, this information should be included in the Engineering Product Specification as additional data.

1.6.6 AC Output Current Devices: For devices that output AC power, typical AC input specification values include no external AC loading. Maximum AC input specification values include maximum resistive load ing on the AC output, as well as maximum DC loading if applicable.

1.6.7 Non-Applicable Parameters: If a table calls for data that is not applicable or has no reference to a product being tested, "N/A" should be entered in the table.

1.6.8 Additional Parameters: This section outlines minimum requirements. If additional parameters are required to fully describe the unit in question, add them to the appropriate tables.

1.6.9 Measurement Units: Tables in the appendix must be completed in both SI (metric) and U.S. Customary units.

2. PHYSICAL SPECIFICATIONS   (A New Page)

PARAMETER	DESCRIPTION	SUGGESTED MEASUREMENT ACCURACY
Product Description	A brief functional description of the product (25 words or less). Include drawings of cable access if possible.	N/A
Mounting Code	The following codes can be used: F.S. = Free Standing R.M. = Rack Mounted T.T. = Table Top W.M. = Wall Mounted; MOD = Module	N/A
Height	Distance from lowest to the highest or uppermost point.	±1%
Width	Distance between sides.	±1%
Minimum Width	Minimum width of cabinets with end panels removed.	±1%
Depth	Distance from front to back.	±1%
Weight	Measured weight. Typical: (without options). Maximum: (with full complement of options).	±3%
Shipping Height	Distance from base to top of carton.	±1%
Shipping Width	Distance between sides of carton.	±1%
Shipping Depth	Distance from front to back of carton.	±1%
Shipping Weight	Typical packaged shipping weight.	±5%
Point Load	Maximum force exerted on floor at points (i.e., wheels or feet) of contact, under worst case loading conditions.	±5%
Shipping Code	The following codes are used: SK - Mounted on pallet or another device. BX - Packed in corrugated carton. CS - Shipped on cushioning casters.	N/A
Minimum Service Clearance	Minimum clearance required for proper airflow & service access to the front, rear, & sides of the device.	N/A
Data Cable(s)	Data cable(s) (type & length) if included with the product.	N/A

3   ENVIRONMENTAL SPECIFICATIONS   (A New Page)

PARAMETER	DESCRIPTION	SUGGESTED MEASUREMENT ACCURACY
Temperature (Operating)	Minimum & maximum air intake temperatures while operating.	$\pm2^0$ F $\pm1^0$ C
Temperature Rate of Change (Operating)	Maximum air intake temperature rate of change while operating.	$\pm5\%$
Temperature (Non-Operating)	Minimum & maximum non-operating temperatures.	$\pm2^0$ F $\pm1^0$ C
Relative Humidity (Operating)	Minimum & maximum intake air relative humidity while operating	$\pm3\%$ RH
Relative Humidity Rate of Change (Operating)	Maximum rate of change of relative humidity while operating.	$\pm8\%$
Relative Humidity (Non-Operating)	Minimum & maximum storage relative humidity.	$\pm3\%$ RH
Maximum Wet Bulb Temperature	Maximum tested wet bulb temperature while operating.	$\pm2^0$ F $\pm1^0$ C
Minimum Dew-point Temperature	Minimum tested Dew-point while operating	$\pm2^0$ F $\pm1^0$ C
Heat Dissipation	Total heat dissipation of product, at typical (for this product only) & maximum (full DC loading,) using nominal line voltage.	$\pm2\%$
Air flow **Note:** Air flow parameter does not apply to free convection cooled units.	Volumetric air-flow rate used by the device cooling fans (typical) at nominal line voltage. If unit can be operated at 50 Hz: use 50 Hz air-flow.	$\pm5\%$
Air Flow Intake & Exhaust Location	Indicate location of cooling air intake & exhaust ports.	N/A

3   ENVIRONMENTAL SPECIFICATIONS (Continued)

PARAMETER	DESCRIPTION	SUGGESTED MEASUREMENT ACCURACY
Air Quality **Note:** If not a product constraint, this does not apply	Air quality requirement that applies to this product includes the minimum particle size included in a count & the maximum number of particles per air volume in that count.	±5%
Altitude (Operating)	Maximum altitude; (or simulated altitude) while operating.	±5%
Altitude (Non-Operating)	Maximum altitude; (or simulated altitude) while not operating.	±5%
Mechanical Shock (Operating)	Shock level (Gs peak) & duration (ms) while operating.	±10%
Vibration (Operating)	Frequency range & vibration level while operating. (Levels typically in Gs peak or double amplitude).	±5%
Mechanical Shock (Non-Operating)	Shock level (Gs peak) & duration (ms) while not operating & packaged for shipment.	±10%
Vibration (Non-Operating)	Frequency range & vibration level while not operating & packaged for shipment. (Levels typically in Gs peak or double amplitude).	±5%
Acoustics	Acoustic level emanated by this product usually measured dBA at certain locations with respect to the product. (Include measurement, data, location, & units, if appropriate for the product.)	±3dBA

4   AC POWER SPECIFICATIONS   (A New Page)

PARAMETER	DESCRIPTION	SUGGESTED MEASUREMENT ACCURACY
Voltage Nominal	Nominal AC voltage nameplate rating for this product. One nominal per AC power specification table only. Additional voltage nominals require additional sheets.	N/A
Voltage Range	Minimum & maximum true rms AC line voltage	±2%
Frequency Nominal	AC line nameplate frequency for this product at the voltage (nominal) specified above.	N/A
Frequency Range	Minimum & maximum line frequency for this product. Testing is required for frequency sensitive products (disks, CD Drives, etc.).	±1/2 Hz
Number of Phases	Number of AC power phases that this product uses.	N/A
RMS Current (Steady State)	True rms current for the option operating at nominal voltage at typical load (this product only, no expansion) & maximum (full-loading) conditions. Record current for all phases, neutral, & ground. **Note 1:** Ground current measurement to be obtained from product safety ground test at nominal voltage. **Note 2:** A true rms responding instrument should be used for this measurement	±2%
Fuse or Circuit Breaker Rating	Rating of the input power AC current protection in the device.	N/A
Power Factor	Ratio of total input watts to total input volt amperes for the basic option (typical-no expansion) & at full power supply output – maximum DC loading conditions.	N/A

4   AC POWER SPECIFICATIONS (Continued)

PARAMETER	DESCRIPTION	SUGGESTED MEASUREMENT ACCURACY
Crest Factor	Ratio of peak steady-state current to true rms current values for worst case phase. For typical product (no expansion) & at maximum DC loading conditions.	N/A
Current Distortion Factor	Total harmonic distortion of input AC line current for this product (typical-no expansion,) during worst case phase. **Note: 3** Include harmonic current information in Engineering Specification if available.	±10%
AC Power Output	List the voltage & maximum current for each output AC circuit. Also give output method & receptacle or connector designation.	N/A
Power Controller Type	If a specific power controller is supplied with this product, indicate type.	N/A
Peak Current (Steady State)	Peak current for the option operating at nominal voltage, at typical load. (this product only, no expansion) & at maximum (full loading) conditions. Record current for all phases, & neutral. **Note: 4** For products that use varying levels of current, minimum current is the minimum operating condition (e.g., CD Drives unloaded, disk not spinning). Typical current is the normal operating current (e.g., CD Drives disk reading or writing operations). Maximum current is the maximum produced during any operation (e.g., head seek).	±4%
DC Current on AC Lines	Typical, (this product only-no expansion) phase, & neutral DC currents.	±2%

4   AC POWER SPECIFICATIONS (Continued)

PARAMETER	DESCRIPTION	SUGGESTED MEASUREMENT ACCURACY
Power Cord Type	Type (attached, number of conductors, type of cord set, etc.,) of line cord supplied with this option.	N/A
Power Cord Length	Length of the line cord supplied with this option.	±5%
AC Plug Type	NEMA or other industry designation for the power plug supplied with the unit.	N/A
Ride-through Time	Dropout or interrupt tolerance for the worst case phase. Maximum time for voltage to drop from the low line condition to 0 volts during operation & return to low line condition without disrupting operation of the product. (Without causing ACLO, DCLO or other interrupt conditions.) Test is to be made at maximum DC loading conditions.	±5%
Inrush Current	Maximum peak inrush current produced by this product on the worst case phase.	±5%
Start-up Current Amplitude	RMS starting current value (after the peak inrush) produced, on the worst case phase, on the first cycle after inrush at the nominal line voltage. Measurement to be made at maximum (full-loading) conditions.	±5%
Start-up Current Duration	Time required for the start-up current to fall to 110% of steady-state rms current.	±10%
Power Consumption	AC input power (watts) used by this product at typical load (this product only, no expansion) & maximum (full loading) conditions.	±4%

4 AC POWER SPECIFICATIONS (Continued)

PARAMETER	DESCRIPTION	SUGGESTED MEASUREMENT ACCURACY
Apparent Power	Total calculated volt amperes drawn by this product. Typical value is the product of nominal volts & typical rms current. Maximum value is the product of nominal volts & maximum rms current.	N/A
DC Output-Watts Available	Total calculated watts of DC available to power additional options (without adding more power supplies). This is the maximum DC output power of the power supplies minus the typical DC load which exists in this product without expansion. **Note: 5** For power supplies, use the maximum specified DC output wattage. Does not apply to products without expansion capability.	N/A
DC Output Amperes Available at Each DC Voltage	Available current to power additional options at each DC voltage. This is the maximum total output current at each DC voltage minus the typical DC current load which exists in this product without expansion.	±5%

5   DC INPUT POWER SPECIFICATIONS   (A New Page)

PARAMETER	DESCRIPTION	SUGGESTED MEASUREMENT ACCURACY
Power Consumption	DC power used by the device in intended use. Typical load, (this product only-no expansion).	±2%
Current at each DC voltage	Current drawn at each DC voltage. Typical load, (this product only-no expansion); (basic option only).	±2%
Voltage Ripple	Maximum peak-to-peak ripple allowed at each DC voltage nominal.	±5%
DC Voltage Range	Minimum & maximum voltage while operating, for each DC voltage nominal.	±1%

6   ELECTROMAGNETIC COMPATIBILITY (EMC)
      SPECIFICATIONS   (A New Page)

PARAMETER	DESCRIPTION	SUGGESTED MEASUREMENT ACCURACY
Field Strength Susceptibility (Operating)	Field strength & frequency ranges while operating.	±3 dB
ESD Level	Electrostatic discharge levels this product passed.	±1 kV

7   REGULATORY AGENCY COMPLIANCE   (A New Page)

This section outlines the regulatory agency compliance that the product meets. There is no table associated with this section.

List the applicable compliance information for each regulatory agency (e.g., UL 478, FCC Class B, C SA, VDE, etc.).

EXAMPLES FOLLOW:

The first edition of UL 4200A, Standard for Safety for Products that Incorporate Button or Coin Cell Batteries Using Lithium Technologies

Standard for Rechargeable Batteries for Mobile Telephones (IEEE 1725) and Standard for Rechargeable Batteries for Multi-Cell Computing (IEEE 1625);

Underwriters Laboratories Inc. ("UL") Standard for Safety for Lithium Batteries (UL 1642)

American National Standards Institute/National Electrical Manufacturers Association ("ANSI/NEMA") Safety Standards for Primary, Secondary and Lithium Batteries (ANSI/NEMA C18);

ASTM International ("ASTM") Standard Consumer Safety Specification 10 for Toy Safety (ASTM F963); UL Standard for Household and Commercial Batteries (UL 2054)

UL Standard for Audio, Video, and Similar Electronic Apparatus– Safety Requirements (UL 60065)

Section 7: Sample Engineering Specification

> **Note: The example provided herein is strictly for illustrative purposes only and not intended to be representative of any actual product requirements.**
>
> Your (actual) Engineering Specification will vary, or by co-incidence, match some parts of the illustration. The statements made in this example are not to be interpreted as design guidance or a guarantee of expected Product performance. You should be aware of the risks of adopting Specifications without carefully considering the input from other design-team members, especially before making any final decisions.
>
> You cannot assume that any recommendations made in any part of this document will meet your Product requirements, except by co-incidence and the contents of this document are provided for information/instructional purposes only.

Remember: When you write these documents, you must be thorough, but also concise. For example:

- Always use the approved Corporate "Word-Processing" format, including any special templates, (if available.)
- Always put a date stamp at the top of the pages, (the header) showing when the file was most recently updated.
- Maintain a "revision" page.
- Keep copies of "Old Revisions" in a Project Folder

The second document you should create describes the data that shall be included in the Engineering Specification (focus on Electro-Mechanical Packaging,) to provide a source for consistent and accurate information about (all) [*Insert Your Company Here*] products. The document will permit Engineers/Test Personnel the ability to collate all relevant parameters about a specific product design in one place (the Engineering Specification,) and make this information available to all [*Insert Your Company Here*] personnel who deal with customers. Usually 8 ½ x 11 paper, the document contains:

A Title Page   (ON ITS OWN PAGE)
A Revision History/Status   (ON ITS OWN PAGE)
A Table of Contents   (ON ITS OWN PAGE)
Abbreviations/Definitions   (ON ITS OWN PAGE)

(An "actual" E/M Packaging Specification is reproduced for illustrative purposes in the following pages.)

Title Page

Electro/Mechanical Packaging Specification:
The "Jim-Dandy" VIP Printer Project

Compliance/Engineering Model
(Post Prototype Design "C")

AUTHOR: Your Name Here
Original: 0.0   14 October 2016

Revision History/Status		
Date	**Version**	**Description of Change**
10/14/2016	0.0	Initial Release
10/15/2016	1.0	Added regulatory Agencies to Section 7.0
10/16/2016	2.0	Added Notes and Appendices
…	…	…

TABLE OF CONTENTS

TABLE OF CONTENTS (Continued)

ABBREVIATIONS & DEFINITIONS

Electro/Mechanical Packaging Specification:
The "Jim-Dandy" VIP Printer Project

A.M.D.	Air Moving Device
Db	decibels
DMT	Design Maturity Testing
DVT	Design Verification Testing
ECO	Engineering Change Order(s)
E/M	Electro-Mechanical (Packaging)
FRU	Field Replaceable Unit
I/O	Input / Output (Signals or Cables)
MTBF	Mean Time Between Failures
MTTR	Mean Time to Repair
PC	Printed Circuit (for purposes of this document)
PCB	Printed Circuit Board
PVT	Process (Manufacturing) Verification Testing
RFI/EMI	Radio Freq. Interference/Electro-Magnetic Interference
TTR	Time to Repair

(Add other definitions as required – especially those unique to your product)

Electro/Mechanical Packaging Specification: The "Jim-Dandy" VIP Printer Project

1.0 SCOPE

   1.1 Purpose of This Document:

      This Document will identify the design parameters/criterion, and methods necessary to produce a cost effective Electro-Mechanical Package for the "Jim-Dandy" VIP Printer Project. Its focus is E/M Packaging only, except as required for interfacing.

   1.2 Statement of Revision Policy:

      This Document will be continuously revised as information becomes available, numerical values are computed or changed, and/or the product specifications are altered. Periodically this document will be fully reissued in printed form under a higher revision marker.

   1.3 Exceptions:

      This document will not include the following Engineering Documentation:

- Mechanical Layouts Assembly Drawings
- Electrical/Logic Design Documentation

      Because of the complexity of the "Jim-Dandy" VIP Printer design and its unique stand-alone capabilities, there will be no attempt to specify or set design constraints for the "Jim-Dandy" VIP Printer in this document. It is understood that the FINAL integrated package is what will meet the requirements set forth in this document.

   1.4 Introduction:

      This document describes the Mechanical Design parameters that will be used as design input definition/information for the "Jim-Dandy" VIP Printer Project. It will allow all members of the design team a single place to find information and to state their criteria for their designs. It is the interface document which will tie together the E/M packaging with the product requirements.

2.0 APPLICABLE DOCUMENTS/REFERENCE DOCUMENTS

   The following documents form a part of this specification. Unless otherwise noted the latest revision of each in effect shall apply.

- [Your Company Name Here] Specification: General Quality and Reliability Requirements for [Your Company Name Here] Hardware Products.
- [Your Company Name Here] Procedure: Hi-Pot Testing

3.0 REQUIREMENTS

   These are the minimum qualification requirements to verify design integrity.

   3.1 Environmental:

      The requirements for operating temperature ranges are based on "free-air" operation within temperature ranges normally found within home or commercial sites.

      a. Conventional cooling techniques, compatible with existing environments, are used throughout.

      b. Meet the functional, structural requirements for shipping and handling.

Page 6 of 23

c. No degradation or alteration in performance to adjacent equipment or installed equipment because of a particular design approach.

Note: Exceptions to [Your Company Name Here] Specifications that have been made and are noted as applicable.

3.1.1 The Normal Operating Temperature for The Unit Is:
+5 Degrees C/+41 Degrees F to +35 Degrees C/+95 Degrees F. (Corresponds to: [Your Company Here] specification changes as noted). The limiting factor for high temperature operations is the survivability of the media. The maximum allowable temperature is based on operation at sea level, i.e.760mm Hg (29.92 in. Hg); the maximum allowable temperature will be reduced by a factor of 1 Degree C per 1,610 M (1.8 Degrees F. per 5,280 Ft., for operation at higher altitudes. The operating humidity range for the unit is: 25% Maximum @ 95 F, then 40% @ 90 F to 70% @ 85 F to 95% @ 41 F, with a maximum wet bulb temperature of 32 C (90 F,) non-condensing throughout the range. This is to compensate for the "Apparent Temperature" effect.[8]

3.1.2 The non-operating (shipping/storage) temperature range for the unit is: −23 Degrees C (−10 Degrees F) to +60 Degrees C (+140 Degrees F).

NOTE: These are the limits without the media installed and with the product shipping material in place for protection. The non-operating Humidity range for the unit is the same as the operating humidity range with these additions: (non-condensing throughout the range) 25% Maximum @ 95 F decreasing to 10% Minimum @140 F, and 95% Maximum from −10 F through +40 F.

a. Temperature cycling shall be gauged at ±10 Degrees C (18 Degrees F) per hour, increase or decrease during the storage cycle. Stabilization at each value will occur for 1 hour. Two complete cycles are the requirement with no degradation of performance.

b. Temperature cycling shall be gauged at ±28 Degrees C (50 Degrees F) per hour, increase or decrease during the transportation cycle. Stabilization at each value will occur for 2 hours before increasing or decreasing temperatures. This unique condition is due to situations encountered in air or truck transport and variable weather conditions. Two complete cycles are the requirement with no degradation of performance.

c. Accelerated aging will be performed per [Your Company Name & Specification] using both conditions of moisture.

Page 7 of 23

---

[8] Apparent temperature is the temperature equivalent perceived by humans, caused by the combined effects of air temperature, relative humidity and wind speed. The heat index and "humidex" measure the effect of humidity on the perception of temperature.

      d. Temperature shock testing [Your Company Name & Specification] is NOT a requirement for this product.

3.1.3 Operating Shock & Vibration [Your Company Name & Specification] The unit shall withstand the following with no degradation of performance:

- A six-inch flat bottom, free-fall drop (No Packing Material).
- A three-inch corner drop, opposite corner supported or resting on plane surface, all four corners (No Packing Material).
- A 20g impulse of 63cps = 0.10-inch displacement. On the following surfaces: front, each side, rear & top (simulating an accidental or intentional "hit" while operating – the unit shall continue to operate.
- There is no requirement for operating vibration at this time

3.1.4 Non-Operating Shock & Vibration:

      a. Non-operating shock (with all shipping materials/cartons in place): **Note:** ALTHOUGH THE SHIPPING MATERIAL MAY BE CRUSHED DR DAMAGED, THE UNIT WILL OPERATE WHEN REMOVED FROM ITS BOX.

      Individual Carton
- A 36-inch free fall drop on each of 6 axis
- A 36-inch free fall drop on 2 opposite corners.
- With the units stacked on a skid or in their shipping container(s,) a 36-inch free fall, flat-bottom drop.

      b. Non-operating Vibration testing shall simulate truck transport of palletized units. The vibration shall be RANDOM with a range of 0 to +25gs, 10 cps to 1000 cps, inclusive. The test shall run for one hour.
- Additional vibration (as required)

3.1.5 Altitude Requirements:

The units shall be capable of OPERATING from: sea level (0 Ft) to +10,000 Ft (3,050 Ml. The units shall be capable of TRANSPORT (non-operating) from: sea level (0 Ft) to +40,000 Ft (12,200 M) and return to sea level with no adverse effects. This requirement is due to Air Transport.

3.1.6 Sand & Dust per: [Your Company Name & Specification]

3.1.7 Fungus Resistance per: [Your Company Name & Specification]

3.1.8 Corrosion Resistance:

The Salt Spray test in [Your Company Name & Specification] may be used to verify the corrosion resistance of the enclosure. However, the electronic components, PC Boards and connectors are not intended to operate after this test. At the Commercial Quality grade of these components there will be severe corrosion effects and possible electrical shock hazard. It is imperative that no E/M components be operated during this particular testing phase.

3.1.9 Cycling Power ON/OFF:

A major portion of electronic equipment failures occurs during the power on/off/on cycle. There are several series of events which may cause a failure. For the purposes of this design we will consider the following:

a. cold start (all components at ambient temperature)

b. warm start (system used within the prior 60 minutes)

c. hot start or rapid on/off caused by power loss or fluctuation or rapid switching after unit has been operating for 60 mins.

In addition, line voltage variations (either high or low) significantly affect the survivability rate. The design will meet the following criterion to demonstrate its reliability

Conditions	Line Voltage @ −15%	Line Voltage @ "Nominal"	Line Voltage @ +5%
Cold Start	ON − 5 minutes	SAME PROCESS 9 Repetitions	SAME PROCESS 3 Repetitions
	OFF − 20 minutes		
	ON − (record result)		
	3 Repetitions		
Warm Start (System Has Been Operating for 60 Minutes. Then OFF: for 30 Minutes)	ON − 10 minutes	SAME PROCESS 9 Repetitions	SAME PROCESS 3 Repetitions
	OFF − 10 minutes		
	ON: (record result)		
	3 REPETITIONS		
Hot Start (System Has Been Operating for > 60 Minutes)	OFF − 3 seconds	SAME PROCESS 9 Repetitions	SAME PROCESS 3 Repetitions
	ON − 1 minute		
	OFF − 3 seconds		
	ON − (record result)		
	3 REPETITIONS		

This testing may be performed in conjunction with burn-in testing.

3.2 Performance/Availability:

The final units, as shipped from Manufacturing, are required to meet:

a. Specified TTR, MTTR per 5.3

b. Specified MTBF per 5.2

c. Regular, in process verification of these attributes

3.3 Manufacturing/Mobility:

The units are required to demonstrate a design that will offer:

    a. Low life-cycle cost
- High Reliability = Low Failures
- Low Cost to Repair

    b. Ease of assembly
- Low skill levels required
- Transition to Robotics assembly

    c. Ease of testing & verification in process

    d. Ease of shipping & handling

    e. Ease of customer, on-site, integration

3.4 Independent Product Certification:

In order to comply with universal certification requirements, the unit shall be designed to the following standards:

- UL-122 Photographic Equipment
- UL-478 Electronic Data Processing Units & Systems
- HEW 21 CFR, Part 278, Subpart C – Radiation Limits
- CSA – C22.2, No. 154-1975 Data Processing Equipment
- IEC-435 Safety of Data Processing Equipment
- IEC-380 Safety of Data Processing Office Machines
- FCC, Part 15, Subpart J, Class A EMI Requirements and Testing of Computer Devices
- VDE-0871/6.78 RFI Limits Equipment Operation Above 10 KHz
- RFI limits Equipment Operation Below 10 KHz
- TUV, per IEC-435 and IEC-380 Energized Office Equipment

4.0 CONFIGURATIONS

4.1 Overall Size & Weight:

Size: 30.5cm x 30.5cm x 8.25cm

    12.0 inches' x 12.0 inches' x 3.25 inches

Weight: (approximate)

Print Engine	3.75 pounds
Electronics	4.50 pounds
Enclosure	3.00 pounds
Total	11.25 pounds

4.2 Features:

Since the bulk of the design will consist of adapting existing technology to the package, and/or mating to existing installed customer equipment, special features concerning advanced design techniques are not anticipated. There will be increased emphasis on FRUs and customer service ability features in the various project requirements.

4.3 Restrictions:

The mechanical interface will be designed to accept the following inputs ONLY: S-VIDEO RGB HI-RES (also Y, R-Y, B-Y) NTSC for: Phillips, SONY, Canon, VCR, Fuji-x Computer. Modifications of the basic design parameters to add options not meeting these requirements will be considered on a cost-plus basis.

5.0 DESIGN & CONSTRUCTION

5.1 General:

The basic options will be housed in a plastic, injection molded enclosure, made of fire-retardant material, (94V-O,) of sufficient strength to withstand shipping, handling and normal use. The enclosure will be covered by a sheet metal top which is used to provide RFI/EMI integrity. The design and placement of all components will adhere to the workmanship standards previously referenced, and the dictates of serviceability and reliability.

The exact order of precedence for design and/or assembly of all components shall be as follows:

- FIRST: Engineering Specifications
- SECOND: Assembly Drawings, Notes or Documented Instructions
- THIRD: Manufacturing Process Sheets
- FOURTH: Verbal directions later implemented via ECOs in the above

The following considerations are GENERAL GUIDELINES to be used for deciding design trade-off during mechanical analysis:

a. Physical Attributes
   i. Should it be a radical departure from current trends
   ii. Different outlines for different markets
   iii. Different outlines for different levels of functionality
b. Environmental Operating Standards
c. Base System Requirements vs. Expandability
   i. Is the entry level system non-expandable
   ii. What are the definitions of an expandable system
   iii. Are all levels of functionality required to have the same form factor
   iv. What is the largest application envisioned
   v. Could an advanced technology satisfy cost goals
d. Communications Requirements
   i. With existing systems
   ii. With remote devices
e. Serviceability Goals
f. Reliability Goals
g. Manufacturing Cost Goals
h. Cost Goals vs. Mark-ups

5.2 Reliability:

The sum-total reliability of the unit shall be: >50,000 hours (electronic) and >10,000 cycles. Reliability shall be calculated using MIL HDBK-217 criteria and tested using standard DMT and PMT methods. DVT will be used as an on-going confidence demonstration. Single-point failures caused by E/M interconnects will be carefully analyzed, and high quality components will be specified to negate or limit the possibility of a mechanical interconnect failure contributing to a system failure.

The greatest failure-mode risk is from the cycling of the electronics off and on. Particular attention must be placed in the "front-end" power supply design to guard against this.

5.3 Serviceability(TTR):

Note: This is explicitly for FRUs and not field-repair of power supply or other major components.

The measure of serviceability is: TTR (Time to Repair). For this design the following will be met:

1. Time to R & R logic modules after diagnosis, 2 minutes' maximum, including removal of access covers.
2. Time to R & R an external data cable, < 1 minute.
3. Time to R & Ra power supply module, (excluding the mother board,) (15 minutes. Including removal of access covers, and reconnection to full operational status.
    4. Time to R & R the mother board, <15 minutes. Including removal of access covers, fasteners option cards, and reconnection to working
    5. Time to R & R Air Moving Device, <10 minutes. Including removal of access cover, connectors and reconnection of same.

Only valid if the "final design incorporates an A.M.D.

6. Time to R & R Printer, (30 minutes. Including access cover removal, cabling and reinstallation.

These goals only apply to new design packaging under our control. They do not include the additional time required to repair individual piece parts. All times are based on one person affecting the repair with sufficient documentation available and prior training/skill to affect the repair in the times indicated.

The MTTR (Mean Time to Repair) is the average time to detect a 100% repair after diagnosis. The MTTR for this design is <13 minutes, and it does not include off-line repair of parts.

5.4 Availability:

Availability is the percentage of time during which the equipment is capable of performing its function. Except for the time the equipment is actually printing the availability will be = 100%. There will be no PM aside from roller cleaning planned for this design.

5.5 Cables/Connectors General:

(see 5.7.5 for specific applications) The unit will interface with external products for its basic inputs. The interfaces will be made over a variety of cables depending on the type of external equipment and the desired input signals.

5.5.1 External I/O:

Analog and video signals are as follows:

−Monitor Output −NTSC −RGB −S-video −Hi-Resolution

5.5.2 Internal Cables/Connectors:

- Wiring requiring manual operations will be discouraged during the design process.
- Internal connections will be accomplished in etch wherever possible
- Connectors will be PC Board mount and wave-solderable.

5.6 Other Cable/Harness Management:

When no viable alternative exists, a hand-made cable or wire harness will be used. This method is discouraged because of the cost involved and labor required. In the event a harness is needed, every attempt should be made to have the harness a "one-piece" assembly rather than individual wire strands. Sufficient service loop material shall be built into each cable to allow easy part removal and replacement. Cables adjacent to mechanisms shall be secured to prevent their entanglement. If Cables must "ride" with an assembly, they shall be self-coiling and secured at both ends.

5.7 Backplanes & Modules:

The design under consideration would entail the use of a single large board to contain all the control logic. In the event the number of components required to execute the design exceeds this board's capacity the alternate approach is to split the logic circuitry into an analog and digital section, each on its own board. These single boards would plug into the large board and we would have a "mother" – "daughter" board arrangement. In either design, the mother board would contain all the connectors necessary to interface the control panel card, the power supply card, the "Jim-Dandy" VIP Printer engine, the I/O interconnect card and any number of additional analog or digital logic cards.

Note: The terms "Backplane" and "Mother-board can be used interchangeably throughout this discussion to denote a particular PC Board(s) design approach in which all multiple interfaces to occur on a single card.

5.7.1 Mother Board:

The mother board will contain all the interconnects to the print engine, the I/O's for external control signals, the power supply card & system power distribution (in etch,) and the front control panel card. It will also distribute signal buses and isolate grounding as required. Basic: construction will be:

- 4 layers (minimum) as follows:
  - o Layer 1 TOP – Signal
  - o Layer 2   – Grounds
  - o Layer 3   – Power
  - o Layer 4   – Signal
- 0.062 ± 0.005 inches thick 94V-0 material (May need to be impedance matched).
- 2-ounce copper on power and ground planes
- Plated-thru hole

- The area directly under the power board will contain no etch lines on Layer #1.

5.7.2 Power Supply Board/Power Supply Section

The need to separate the power components from system elements is vital to: facilitate repair of the probable high failure rate components in the system, to reduce the entry cost of the mother board, reduce system repair time and fault isolation, and to facilitate testing in Manufacturing. The power board will contain the AC Input Section Powerline filters, conversion (AC to DC) electronics, and distribution connectors. The board will mount on a power supply plate using common fasteners and use high contact pressure connectors for interface.

Basic construction will be:

- o 2 layers as follows
- o Layer 1 (TOP) = ground/ground returns
- o Layer 2 (BOT) = Power
- 0.062 ±0.005 inches thick 94V-0 material
- 6-ounce copper
- plated thru holes

Note: The term "high failure rate component" is relative. Power Supplies are the suspect failure item in any system design due to cyclic effects.

As part of the AC power cord set/input on the power board, there should be a power entry "module" EMI filter. This filter component should include the following:

- IEC style connector
- Fuse Holder (Location for Spare Fuse incorporated)
- Voltage selection switch
- Safety interlock prevents fuse removal with line cord inserted
- UL Recognized, CSA Certified, VDE Approved

5.7.3 Daughter Cards:

For purposes of this design, the following cards (in addition to the power supply card) are considered "Daughter Cards" because they plug in-to or connect and communicate through the Mother Board

- I/O Interconnect Card
- Control Panel Card
- (If required)
- Analog Card
- Digital Logic Card

Basic Construction will be as follows

- 2 to 4 layers of etch
- 0.062 ±0.005 inches thick, 94V-0 material
- 2-ounce copper
- Plated-thru holes

The daughter cards will mount vertical to the mother board.

5.7.4 Control Interface Circuit Board:

The control Interface Circuit Board will mount directly to the front panel assembly. It will hold the control switches and illuminated status indicators.

Basic construction will be as follows:

- 2 Layers
    - Layer 1 (TOP) = Signal
    - Layer 2 (BOT) = Power & Ground
- 0.062 ±0.005 inches thick 94V-0 material
- 2-ounce copper
- plated thru holes

The interconnect between the control interface card and the mother board will be via a flexible harness assembly because of the assembly process and tolerance conditions anticipated.

5.7.5 Connectors:

The connectors used to interface the daughter cards (Analog/Digital if required) will be top entry edge-card type. In order to accommodate power distribution requirements to ALL daughter cards and the "Jim-Dandy" VIP Printer engine, each contact used for power will be rated at >1 AMP (@ 5 VDC,) and if necessary, multiple connector pins shall be used for power distribution. The ratio of Ground Return or Ground Reference pins to power pins shall be no less than 1.5 ground to each power. Connectors used exclusively for power shall contain 20% reserved as spare pins, but in no case less than two pins spare. Spacing on power connector pins shall be > 0.125 in. with 0.156 in. preferred. This is to allow etch of sufficient current carrying capacity and safety spacing to pass across connectors. Signal connectors from mother board to daughter boards will be 0.100 in. pitch per contact and these connectors will contain 10%, but no less than 4 spare pins per connector. The ratio of ground pins to signal pins shall be 1 ground pin per 5 signal pins (Minimum) with special signal requirements, e.g. clock lines needing more. Connectors to the "Jim-Dandy" VIP Printer engine will be PC Board mount, top entry, to accept either mating flexible circuitry or connectorized harness assemblies. They shall be female on the mother board. Minimum spacing shall be 0.125 in. for signal connectors and 0.156 in. for power. The percentages for spare pins will be consistent with other system connectors.

5.8 Logic Board Enclosure Design & Construction:

The base of the enclosure will be a 94V-0 plastic which will be conductively coated on its interior surfaces to aid in RFI/EMI integrity. The conductive coating will be a permanent, non-migrating type which will provide a 0 ohms' path for all emissions to ground. The base will include all mounting bosses and ventilation openings required in the final product.

To insure proper cooling and guarantee RFI/EMI integrity, the design will incorporate a perforated steel plate in the base, (steel is chosen rather than aluminum because of its superior electromagnetic absorption qualities). This plate will be ground connected to the conducted coated surface to maintain emission suppression integrity.

5.8.1 Module Spacing:

If the alternate approach of separating the logic into two modules is required, they shall be placed 1.00 inches apart (on centerline) The modules shall be spaced no closer than 0.750 inches from the metal RFI wall between the electronics bay and the printer cay. That module will have the etch side towards the shield, component side facing away from the Print Engine. The adjacent module will also have the component side away from the Print Engine.

5.8.2 Rear Connector Card:

The Connector card will contain PC Board mounted I/O connectors for all the interfaces required. It will be mechanically fastened to the rear wall of the enclosure after insertion into the mother board. It must be supported securely at each connector location either by locking hardware on the connector or additional security

5.9 Built-In Maintenance Features

The entire mechanical design will be geared towards ease of service, especially by non-technical users/customers. To achieve this, the design direction will be to utilize as many "snap-apart" assemblies as possible. The intent will be to simplify the design and provide obvious visual aids (using labels and pictorial instructions,) for repair.

5.10 Cooling

The entire unit is considered as a single "thermal-entity" in that all techniques and devices necessary to maintain the component junction temperatures during operation will be integral to the design and not supplied externally. To assure adequate cooling the following are guidelines:

- operation of the unit is based on a range of ambient temperatures and relative humidities consistent within the expected marketplace.
- the plot of Apparent Temperatures is used
- consideration of two separate issues:
  a. internal temperature rise, heat generation and dissipation
  b. external heat sources which may influence operation

Based on a system input of 50 watts (Maximum) and considering worst case dissipation factors, the following limits are set (necessitated by media reliability)

- 95 Degrees F is the maximum internal enclosure temperature "Film critical"
- Rate of change per printed circuit board, assuming uniform heating or cooling, Btu input constant, is expected to be: 4 Degrees F (2.2 Degrees C) per minute. The question of "How long until we reach 95 Degrees F internally is not answerable with a standard number. The generalized equation is: $T_s + [(4N) (x)] J = 95$ Degrees F.

Where: T_s = Temperature of ambient air at start. x = Number of minutes to reach the difference at the rate of 4 Degrees F times the number of boards: (N) in active use. (**Note:** if T_s = 95 F then x = 0). Several approaches may be taken to prevent the interior temperature from reaching and staying at 95 F.

    a. Insulate/Isolate the printer engine compartment from all external heat sources as follows:

- Insulate interior walls to absorb heat flow and direct it to the metal cover to provide wide area dissipation
- Attach a "Heat Shield" with an air gap under the unit to shunt heat away from the "Jim-Dandy" VIP Printer compartment which may come from external sources, (sitting above another piece of equipment). The heat shield may extend to the rear or side of the unit and combine with an external heat sink.

    b. Use "cold-plates" or thermal mass transfer techniques to maintain even temperatures. The issue here is one of cost and the need for large areas to successfully accomplish the heat transfer. This is a viable alternative to using a fan or Air Moving Device.

    c. Use an Air Moving Device to circulate cool air through the system. This solution will not prevent the unit from reaching its maximum temperature, but it will delay that point. For example: if ΔT of the unit is kept: < 20 Degrees F. (11Degrees C,) then the operation with T_s(ambient) $\leq$ 90 Degrees F) can be maintained.

Note: ΔT is the difference between the air inside (or leaving) the unit (in forced-air cooling,) vs. the air outside (or entering) the box. Testing mock-ups is one way to determine the success of the final approach.

    5.11 Power Input Design:

The power section will contain the following elements:

- AC Line Filter/Safety Device
- Chassis Ground Stud
- 3 wire AC Input (Brown, Blue, Green with Yellow Tracer)
- All connections to be direct Solder or pressure fit contact
- Customer switchable for 115VAC/230VAC operation
- Chassis ground will be per UL requirements

5.11.1 Power Specifications

Input Voltage	115/230 V AC (±25 V AC)
Current rating	TBD Amps @ 115 V
	TBD Amps @ 230 V
Input Current:	TBD Amps @ 115 V
	TBD Amps @ 230 V
Inrush Current:	TBD Amps Peak for ½ Cycle
	@ TBD Volts Rms
Safety Device (Fuse Rating)	TBD Amps @ 115 V
	TBD Amps @ 230 V
Apparent Power	TBD Volt/Amps

Power factor the ratio of input power to apparent power shall be > 0.65 at full load and low input voltage.

Output: +5 VDC @ TBD Amps
Output: −5 VDC @ TBD Amps
Output: +12 VDC @ TBD Amps
Output: −12 VDC @ TBD Amps

(Other voltages may be derived locally from those supplied)
Ride through (Minimum) after power interruption: TBD
5.12 RFI /EMI:

The major operating clock speed is 20 MHz. In order to compensate for any "micro-environmental" effects caused by the close proximity of this unit to other RFI/EMI sources, (because all equipment is susceptible as a transmitter or receiver,) and to consider harmonic effects below the 0.25(quarter) wavelength, the expected attenuation goal is 60dB for 0.01 wavelength or 0.59 in. (0.15m). This corresponds to contact points to ground of 1.00 inch or less with a 0.25-inch overlap of a free space joint of 0.113-inch gap. Openings shall be ≤ 0.28 sq.in. (circular) and the largest unbroken entry dimension will be ≤ 0.59 in. In order to achieve greater opening sizes, the use of perforated steel sheet stock is recommended, and metal to metal bonding techniques used throughout. All clock lines will be treated as co-axial lines and laid out with the shortest possible etch runs. They will be impedance controlled, using strip-line and adjacent ground runs. The clock lines will cross power and signal bus runs at perpendicular angles to reduce crosstalk and/or coupling effects. The use of DC motors in the "Jim-Dandy" VIP Printer requires that power for the motors be isolated from any logic power runs, to prevent conducted noise problems and that the motor returns or ground lines be isolated from logic and control ground lines. All grounds/returns will be tied back to a single-point ground at the power supply module. The top cover of the unit will be of sheet steel stock. It will be conductively coated on its interior surfaces to provide ground bonding with other elements of the system.

There will be three specific grounding locations in the unit which will directly mate with the top cover via a series of compressible BeCu finger stock attachments.

- The top of the power supply mounting plate
- The perforated sheet stock where it rises adjacent to the Print Engine to form a wall.
- The rear interconnect printed circuit board top surface, (The rear board will also form a ground connection to the interior sheet-stock assembly.)

Additional grounding points will be determined after preliminary RFI/EMI scan testing.

5.12.1 Design Guidelines for RFI/EMI Control:

The suppression of RFI/EMI emissions is a multi-faceted task. It cannot be accomplished by relying on one design area (alone) to be the "stopper" for the system. In outline form, these are the basic levels of packaging which contribute to RFI/EMI control.

- Component level
- Circuit Level/Component application
- Printed circuit Board level/ Interconnects
- Power supply distribution
- Packaging

These are arranged as follows:

- LEAST COSTLY/MOST PERMANENT = Component, Circuit Level
- MOST COSTLY/LEAST PERMANENT = Packaging/Enclosure

5.13 Mechanical Operating Switches:

There are a number of mechanical switches which are used externally and internally to control the operation of the unit, or to indicate positions of the film-pack within the Print Engine. They are listed below:

Switch Designation	Location	
S1 Door Position Switch	Film Loading Door	Print Engine
S2 Film Box "Present"	Film Handling Module	Print Engine
S3 Stop "Spread"	Film Handling Module	Print Engine
S4 Start Print	Film Handling Module	Print Engine
S5 End of Print	Slow Scan Assembly	Print Engine
S6 Film Speed	Slow Scan Assembly	Print Engine
S7 Power switch	Outside Front Panel	
S8 Print Switch	Outside Front Panel	
S9 "Counter" Switch	Outside Front Panel	
S10 Film-Box Memory	Film Handling Module	Print Engine

The status-line for Switches within the Print Engine are sent to the microprocessor as program inputs. The switches are common wired to ground on one side with the exceptions of:

- S7 Goes to Power Control Circuit
- S8 Returns to microprocessor
- S9 Goes to counter Display

The system will not operate unless the "door-closed" switch indicates that the film-loading door is closed.

5.14 Electro-Mechanical Packaging of Print Engine Control Components:
The Print Engine contains six (6) major elements:

1. Encoder Assembly – Contains:
   a. Scan Motor
   b. Emitters
   c. Sensors
2. Slow Scan Assembly
3. LED Module
4. Optics Module
5. Film Handling Module – Contains:
   a. Spread Motor
   b. Control Switches
   c. Door Closed Switches
6. Pixel Delay Chip PC Board (May be externally integrated in the final design)

(The Spread Cycle Motor and the Scan Motor both require 12VDC to operate.) Within the Encoder Module are two Emitters and two Sensors. The Emitters require +2.0VDC for operation and the Sensors require +10.0VDC for operation. Their ground connections are common connected and the outputs are connected to the micro-processor board. The Pixel Delay Chip PC Board controls the operation of the LED module and directly interfaces its output signal to the LEDs over four separate lines. The input for the Pixel Delay Chip PC Board comes from the Digital Logic section outside the Print Engine. It contains the Pixel Delay Chip, the Red, Green 1, Green 2, & Blue generators and Current Driver Circuitry. The LED module is an integrated unit which has four input connections and a common ground connection.

5.14 Front Panel Indicators:
The front panel/operator's panel will have the following controls & indicators:

Power Switch (S7)	Mechanical ON/OFF Switch
Video Signal	Lighted Indicator
Film Present	Lighted Indicator (Flashes when Empty)
Number of Copies	Lighted Indicator (Displays Number)
Number of Copies (S9)	Mechanical Switch Stepping
Exposures	Lighted Display Indicates Number of Prints MADE During Current Print Cycle
Print Switch (S8)	Mechanical START/STOP Switch
Print-In-Process	Series of 5 LEDs (Mounted Vertically) Which Light in Sequence During the Print Cycle (Top to Bottom) Indicate the Print Process is Normal

6.0 SAFETY

6.1 Electrical Safety:

All precautions for power line, high voltage, energy hazard and dead metal grounding will be designed in to meet UL, CSA and IEC safety rules in effect at the time of production release.

6.2 Mechanical Safety:

Mechanical safety will be designed-in: especially in operator access areas; by rounding and smoothing corners and edges, using gripping surfaces on any devices or subassemblies that require insertion or-withdrawal, with some positive force applied. There will be no openings in the bottom of the enclosure large enough to allow burning particles to fall to the floor or outside the enclosure confines.

6.3 Operator Safety:

There is a trade-off to be made concerning the need to maintain a user friendly philosophy and protecting the operator and/or devices from damage or injury. Those areas necessary for access will be strictly controlled and partitioned from areas of danger.

7.0 ENVIRONMENTAL TESTING

In addition to the series of tests performed to verify the requirements in Section 3.0 of this document the following test will also be performed.

7.1 Electro-Static Discharge:

Special consideration will be given to the problems of electro-static discharge caused by an operator (charged,) touching the unit at a relatively different potential (uncharged). All external to internal parts will have ground points established to safely drain the discharge through to the unit ground.

a) Electro-static discharge testing will be done during environmental testing, on operating units.

b) A static discharge device will be used to simulate human body discharge to another object.

c) The following points will be tested for static-discharge sensitivity:
   1. Printer
   2. External Skins
   3. Access Covers
   4. Connectors/Connector Panel
   5. Operator's Display
   6. Other Areas

The unit will withstand an electrostatic discharge of up to 5 kV with no component destruction, and up to 3 kV with no component malfunction. ESD testing shall be performed by the manufacturer with capacitance equal to 100 pF, series resistance equal to 1500 ohms and spark gap equal to 0.1 to 0.2mm. Upon completion of this testing the unit shall meet all performance requirements of this specification.

7.2 Environmental Test Plan:

The unit shall meet or exceed the requirements of all environmental standards required to market it as a product.

An outline of significant tests (additional) follows:

1. Power Line Voltage Variation
2. Over-temperature Stress Testing
3. Temperature/Humidity Tests - Operating/Non-Operating
4. Mechanical Shock/Vibration Tests - Operating/Non-Operating (including free fall flat bottom drop testing)
5. Electro-static Discharge Testing
6. FCC Testing - Emissions/Susceptibility
7. Hi-Pot Testing

During operating portions of the environmental test error log shall be maintained. During the Line Voltage Variation Testing and the over-temperature Stress Testing, Thermo-couple probes will be attached to key components for recording temperature rise. This information will be used to satisfy CSA requirements.

7.3 Configuration of Equipment During Testing:

1. All operating temperature testing shall be performed with the unit configured per maximum power load and customer configuration
2. All non-operating tests will be conducted with the unit in its shipping box.
3. The placement of accelerometers to measure shock inputs and vibration peaks will depend on the final mass configuration of the unit.

8.0 INSTALLATION REQUIREMENTS

8.1 Environmental Conditions:

- The unit is designed to function over a wide range of environmental situations.
- Room temperatures and humidities within the tolerances of the specification can be varied without adverse reliability effects.
- A normal operating range is expected to be tolerated by the operator as well as the equipment, therefore extremes of hot or cold are not the operating norm and not anticipated.

8.2 System Power Requirements:

The prime power source at the user site is to be properly sized for the unit operation. No provisions are made in the design to protect the unit against surge-currents, (e.g. lightning strikes,) peak load build-up, or ungrounded power receptacles. The unit will operate during "brownout" conditions.

8.3 Customer Connection Interface:

The rear connector panel will have plain and clearly defined markings adjacent to each connector. The use of standard BNC connectors will reduce user complexity.

The power cord will be a standard three-wire removable cord set, 36 in. long, or equivalent.

9.0 SHIPPING REQUIREMENTS

Although the product will be a self-contained unit and built to survive the rigors of use and shipping, there are contingencies which must be discussed.

1. During shipping and storage, a reasonable amount of "stacking" of units within their shipping containers will be permitted. Under no conditions should units be stacked without protective support.
2. Transportation and delivery should be done by carriers familiar with electronic equipment handling
3. The units shall be shipped in an upright condition only, and all packing material shall be clearly marked to designate same.
4. Outdoor storage is not permitted under any circumstances.

END OF DOCUMENT

Notice, that by articulating in words your exact approach to solving the problem, it is (almost) possible to visualize the designed product.

It is (also) evident to other members of the design team, "where you're headed" with your design. (Ambiguity leads to increased costs.)

After the specification is completed, the next steps are "multi-tasked" in that the E/M Engineer must "juggle" the following (simultaneously):
- Component Selection & Qualification
- Form/Fit/Functional Layouts
- Testing
- Thermal Analysis (if required)
- Cosmetic/Esthetic Decisions
- Labeling (There is nothing worse than to get to the end of a design; only to discover that there is "no room" for a legally required warning label!)

What does Engineering design documentation (for the final product) consist of?
- Drawings – 3D, 2D, Fabrication & Assembly
- Standard Detail Sheets – with Parts Lists
- Miscellaneous
- Bills of Materials for each Assembly
- Cost Estimates – written, written, written!

Section 8: What could Possibly Go Wrong?

Abstract: A lot of things could go wrong – Some as simple as the product "doesn't work" as advertised and there are returns, to death of the user!

Product Liability *The responsibility of a manufacturer or vendor of goods to compensate for injury caused by defective merchandise that it has provided for sale.*

When individuals are harmed by an unsafe product, they may have a Cause of Action against the persons who designed, manufactured, sold, or furnished that product. In the United States, some consumers have hailed the rapid growth of product liability litigation as an effective tool for Consumer Protection. The law has changed from *caveat emptor* ("let the buyer beware") to Strict Liability for Manufacturing defects that make a product unreasonably dangerous.

Manufacturers and others who distribute and sell goods argue that product liability verdicts have enriched plaintiffs' attorneys and added to the cost of goods sold. Businesses have sought tort reform from state legislatures and Congress in hopes of reducing damage awards that sometimes reach millions of dollars.

Theories of Liability: In the United States, the claims most commonly associated with product liability are negligence, strict liability, Breach of Warranty, and various consumer protection claims. The majority of product liability laws are determined at the state level and vary widely from state to state. Each type of product liability claim requires different elements to be proven to present a successful claim.

Types of Liability: Section 2 of the Restatement (Third) of Torts: Products Liability distinguishes between three major types of product liability claims:
- Manufacturing Defect
  - Manufacturing Defects are those that occur in the Manufacturing process and usually involve poor-quality materials or shoddy workmanship.
- Design Defect
  - Design Defects occur where the product design is inherently dangerous or useless (and hence defective)

no matter how carefully manufactured; this may be demonstrated either by showing that the product fails to satisfy ordinary consumer expectations as to what constitutes a safe product, or that the risks of the product outweigh its benefits.

- Failure to Warn (also known as Marketing defects)
  - o Failure-to-warn defects arise in products that carry inherent nonobvious dangers which could be mitigated through adequate warnings to the user, and these dangers are present regardless of how well the product is manufactured and designed for its intended purpose.

However, in most states, these are not legal claims in and of themselves, but are pleaded in terms of the theories mentioned above. For example, a plaintiff might plead negligent failure to warn or strict liability for defective design.

Breach of Warranty: Warranties are statements by a manufacturer or seller concerning a product during a commercial transaction. Warranty claims commonly require privity between the injured party and the manufacturer or seller; in plain English, this means they must be dealing with each other directly. Breach of Warranty based product liability claims usually focus on one of three types:

1. Breach of an Express Warranty.
2. Breach of an Implied Warranty of merchantability
3. Breach of an Implied Warranty of fitness for a particular purpose.

Express Warranty: claims focus on express statements made by the manufacturer or the seller concerning the product (e.g., "This chainsaw is useful to cut turkeys"). The various implied Warranties cover those expectations common to all products (e.g., that a tool is not unreasonably dangerous when used for its proper purpose,) unless specifically disclaimed by the manufacturer or the seller. Express Warranties are usually statements in writing such as those provided by the manufacturers in owner's manuals and other written sales or advertising materials, or by a sample or model.

Implied Warranties: are broader in scope than Express Warranties and assure consumers that retail products would meet certain minimum standards of quality whereby the product is fit for use for the purpose intended. In each type the manufacturer assumes the liability and responsibility to correct the defect or to repurchase or replace the product.

Typically, the existence, scope and consequence of Express and Implied Warranties is a matter of individual state(s) law, usually covered by Article II of the Uniform Commercial Code.

Lemon Laws: American state laws that provide a remedy for purchasers of cars and other consumer goods, in order to compensate end-users for products that repeatedly fail to meet standards of quality and performance. These may be defective products of all sorts ranging from small electrical appliances to huge pieces of machinery. (However, in a strict interpretation, the term "lemon" is generally thought of as applying to defective vehicles such as automobiles, trucks, SUVs, and motorcycles.) The Federal Lemon Law, (the Magnuson-Moss Warranty Act) was enacted in 1975 and protects citizens of all states. State Lemon Laws vary by state and may not necessarily cover used or leased cars, or other goods. The rights afforded to consumers by Lemon Laws may exceed the Warranties expressed in purchase contracts. (A "Lemon Law" is the common nickname for these laws, but each state has different names for the laws and acts.)

Federal Lemon Laws cover anything mechanical. The Federal Lemon Law also provides that the warrantor may be obligated to pay the prevailing party's attorney in a successful lemon law suit, as do most state Lemon Laws.

> **Note:** There is, also, Strict Liability: Rather than focus on the behavior of the manufacturer (as in negligence,) strict liability claims focus on the product itself. Under strict liability, the manufacturer is liable if the product is defective, even if the manufacturer was not negligent in making that product defective.

Wait Read on!

Product Liability can be both ridiculous and ridiculously costly at the same time:

5 Unbelievable Product Liability Lawsuits – April 8, 2013/3
Comments/in Courtroom /by Lisa [9]

Each year thousands of product liability claims are filed, which eventually lead to a plethora of settlements and verdicts being awarded to consumers who have been injured indirectly or directly by faulty or sub-standard products. These lawsuits, filed by consumers, require companies to maintain safety standards and spend more resources for product testing. Here, we present some notable products that correspond with: *5 Unbelievable Product Liability Lawsuits*.

These cases show how product safety and reliability are crucial.

1. <u>Blitz gas cans</u>: Blitz was the largest producer of portable gas cans in the United States. Headquartered in Miami, Oklahoma, this company filed for bankruptcy in mid-2012 because of a barrage of product liability lawsuits against it. Many consumers from different parts of the country filed cases against the company because the cans would explode when used to pour gas to start a fire. Each of these claims cost the company an average of $4 million and more than 30 cases were filed in 2012 alone, causing the company to close its operations.

2. <u>McDonald's coffee</u>: The Liebeck v. McDonald's case of 1994 is one of the most prominent unbelievable product liability cases in U.S. history. In this case, Stella Liebeck accidentally poured hot coffee, purchased from McDonald's, on her lower body and suffered third degree burns on her thighs, groin and buttocks. Liebeck's lawyers argued that the company served coffee at a temperature of 180 to 190 degrees Fahrenheit while other companies served coffee only at a "reasonable" 140 degrees Fahrenheit. Liebeck was awarded a jury verdict of $2.7 million in punitive damages and $160,000 for medical expenses.

3. <u>Remington rifle models 700 & 710</u>: The Remington rifle models 700 & 710 were proven to have a faulty fire control system, which caused the rifle to fire even when the trigger wasn't pulled. (All that was needed to fire the rifle was the release of the safety latch, which

9 http://www.iveyEngineering.com/unbelievable-product-liability-lawsuits/ 16 October 20

could easily be released when the rifle was simply bumped or jarred.) Many product liability lawsuits have been filed against the Remington Rifle company, but the most prominent jury verdict was $15 million awarded to a Texas man in 1994, when he accidentally shot himself in the right foot while hunting.

4. Ledraplastic balancing ball: In 2009, Francisco Garcia of the Sacramento Kings, was balancing on a 75 centimeter, (30-inch) Ledraplastic balancing ball, along with weights, when the ball burst and he was injured. He fractured his right forearm and was unable to play the first four months in his first-year contract with the Kings. The Kings and Garcia filed a product liability claim against Ledraplastic for $4 million in lost salaries and $29.6 million in damages and eventually won the case.

5. Toyota cars: In 2010, Toyota issued a massive recall for many of its cars. A safety feature known as "brake to idle fail safe" was not installed in many cars and, therefore, increased the chances of an accident when the accelerator malfunctioned. The aim of the fail-safe system is to prompt the engine to ignore the gas pedal when the brakes are pressed, greatly reducing the chances of an accident, even when there is a problem with the accelerator. The failure to include this fail-safe mechanism in many Toyota models resulted in one of the biggest litigation cases in recent history, as well as a class action lawsuit. Toyota agreed to pay a whopping **$1.1 billion** to settle the suit.

– Injury Claims – [10]

Bodily Injury: If an end-user suffers a permanent injury on the job, or by using a piece of "consumer" equipment, they're typically entitled to compensation for the damage to their body and any future lost wages. But, depending on the state, benefits for the same body part can differ dramatically. The case must be proven as negligence on the

---

10 *How Much Is Your Arm Worth Depends On Where You Work*: Each state determines its own workers' compensation benefits, which means workers in neighboring states can end up with dramatically different compensation for identical injuries; by Michael Grabell, ProPublica, & Howard Berkes, NPR March 5, 2015
https://www.propublica.org/article/how-much-is-your-arm-worth-depends-where-you-work

part of the manufacturer, supplier or installer that their "negligence in design or implementation" resulted in the bodily injury to the operator, end-user – assuming certain conditions.

The difficulty with bodily injury negligence is that it requires the plaintiff (the injured party) to prove that the defendant's conduct fell below the relevant standard of care. However, if an entire industry tacitly settles on a somewhat careless standard of conduct (that is, as analyzed from the perspective of a layperson,) then the plaintiff may not be able to recover, even though he or she is severely injured, because although the defendant's conduct caused his or her injuries, such conduct was not negligent in the legal sense (if everyone within the trade would inevitably testify that the defendant's conduct conformed to that of a reasonable tradesperson in such circumstances).

As a practical matter, with the increasing complexity of products, injuries, and medical care (which made many formerly fatal injuries survivable,) it is quite a difficult and expensive task to find and retain good expert witnesses who can establish the standard of care, breach, and causation.

EXAMPLES: Consumers, themselves, were not necessarily physically injured during these particular occurrences – rather the incidents provide the basis for "what could be a tragedy."

Exploding Lap-Tops: Hundreds of millions of lithium-ion lap-top batteries are produced every year, and catastrophic failure, such as explosion or melting, is rare. Still, there have been 43 product recalls for defective lithium-ion batteries since 2002, according to the U.S. Consumer Product Safety Commission.

Batteries can blow-up or melt when internal electrical components short-circuit, when mechanical problems crop up after a fall or an accident, or when batteries are installed incorrectly. But at the heart, all of these failures occur because one portion of the battery gets too hot and can't cool down quickly enough, creating a chain reaction that generates more and more heat – called thermal runaway. During thermal runaway, the miniature battery modules can melt, giving off heat, and the electrolyte material between the anode and the cathode may even boil.

Exploding Cell Phones: [11]

"…The reason you can shove so much power into lithium ion batteries is that lithium basically *"wants to react to almost anything"* – which can lead to explosive results…"

"…one of the most common reasons the batteries can explode is because of mistakes in the charging process. Inside the devices that rely on the batteries there is software that tells them exactly how much the batteries should be charged and how fast. If those protocols are set incorrectly, it can destabilize some chemicals inside the battery and cause a chain reaction that researchers call a *"thermal runaway"* that may lead to fire or explosions…"

"…Another reason could be shoddy Manufacturing or rough user treatment. If unwanted materials, like scraps of metal, accidentally end up inside the battery when it's being made, they can short a cell of the battery and set off a thermal runaway. So could dropping a device if the impact causes a break in the separator between the anode and cathode…"

"…It seems like a Manufacturing problem. Companies report at least 35 cases where the batteries combusted due to *"a very rare Manufacturing process error"* in which the anode and cathode touched…"

"…One company decided to temporarily pull its phone off the market just two weeks after it was released and was offering replacements to people who already purchased the phone…"

"…How often do these types of problems occur…"

"…The good news, is they're pretty uncommon, especially among high-end devices, when manufacturers keep a close eye on production quality…"

---

11 http://www.chicagotribune.com/bluesky/technology/ct-why-those-samsung- batteries-exploded-wp-bsi-20160912-story.html   15 October 2016

"…But there have been plenty of high-profile cases. For instance, back in 2006 Dell recalled more than 4 million laptop battery packs over combustion issues. In 2013, the Boeing 787 Dreamliner was grounded by the FAA, after reports of fires related to the lithium ion batteries used in the planes. And half a million hover-boards, one of the hottest gifts of the last holiday season, were recalled this summer because of lithium ion battery explosions…"

Finally, even kitchen equipment can maim or injure:

... *"Thermomix" product responsible for 45 injuries, says consumer group Choice...* [12]

Key points from the Choice mass incident report:
- ❖ 87 reports of problems with a Thermomix product:
  - o 83 reports of problems for the Model #TM31 product
  - o 4 reports of problems regarding the Model #TM5 product
- ❖ In 45 reports a consumer was harmed:
  - o 18 people reported having to receive treatment from a doctor or nurse
  - o Eight people were hospitalized
    - ▪ Six of those were treated in a specialist burns unit
  - o 12 people were harmed before the October 2014 Model #TM31 recall
  - o 18 people were harmed **after** the recall, but were using a green sealing ring supplied to rectify problems
- ❖ 53 people complained to Thermomix Australia and only five people were happy with the resolution
- ❖ 33 cases of ongoing issues with their machines
- ❖ 26 near-miss cases of "spitting or exploding" hot liquid.

In Summation:

Three difference charges of **Negligence** can be brought against designers in product liability cases:
1. The product was defectively designed (Did not follow commonly accepted Engineering design standards)

12 https://www.theguardian.com/australia-news/2016/may/12/thermomix-products-reponsible-for-45-injuries-says-consumer-group-choice   18 October 2016

2. The design did not include proper safety devices (either, inherent, added-on, or warning)
3. The designer did not foresee possible alternative uses of the product. (e.g. lawn mower for trimming branches)

(In the event you think you can "skate" during a litigation trial – remember the terms "Expert Witness" and "Cross-Examination")

Other charges of negligence (not in control of the designers) are:
1. The product was defectively Manufactured.
2. The product was improperly advertised (used for the wrong intention).
3. The instructions for safe use of the product were not given or understood by the end-user.

(In this scenario "Everyone" in the Company suffers – if the company remains viable after damages are paid.)

Section 9: Drafting

Abstract: "The systematic representation and dimensional specification of mechanical and architectural structures." Drafting personnel design and draw the documents (sometimes referred to as "Blueprints") [13] needed to bring new concepts and products to life. Drafting personnel, also referred to as drafters, prepare a multitude of technical drawings and documentation necessary to build a wide array of products.

Drafting is an ancient art – Dating back centuries. Drafting history goes back to 2000 BC, when a fossilized plan was found showing an aerial view of a Babylonian fortress. Ever since man has been building, erecting or fabricating, man has conveyed technical information about the project. The twenty-first century designer uses his or her technical communication skills by applying an international language of art and graphics to convey information. Drafting uses a strict set of standards and rules, so when a technical picture is created on the drawing board or on CAD, it can be universally understood. A designer communicates ideas, concepts and facts pictorially so that others can manufacture, fabricate, build or construct from these illustrations.

There was a time that a Draftsman actually made drawings using "ink" or pencil! Prior to 3D CAD Drafters made drawings. Yes, we can (still) create what are called 2D drawings today, by creating views or instances of the 3D model and adding dimensions and annotation. These are not drawings; they are what is now called the AID (Associated Information Document).

What then, is an Engineering Drawing?
An Engineering Drawing is a document that describes the part/assembly in an orthographically projected format. The creation of Engineering Drawings can be time-consuming and an Engineer's time is much more valuable than doing grunt design and detailing.

The Engineering Document that is used to convey the information to Manufacturing – is known there as a "Fabrication Document". It is in

---

13 The term "Blueprint" originally referred to a photographic process of printing white lines on blue paper. Later this process was changed to produce blue lines on white paper, (Then called "Bluelines.") Today, the term refers to any photographic print (or reproduction by any means) of plans or technical drawings etc.

a standard format that has been developed over centuries. When done correctly, an Engineering Document "stands alone," without the need for any additional information, explanation or clarity.

There is also a standard procedure for handling these types of drawings. The Draftsman works with an Engineer or designer (or develops the design himself/herself.) He/she would actually do the design utilizing a format called a "layout" (a drawing with no set standards) – then do the Fabrication drawings or pass the layout to other draftsmen to create the piece-part drawings. The drawings are detailed to meet a certain standard. (Even though the Draftsman may have had decades of experience it still has to be checked.)

What is a "Checker?"
A Checker is part of the Drafting-Department internal function. The checker is an experienced Draftsman whose only purpose is to check the drawing. He/she marks every dimension and note with a red or yellow marker. (Nothing is left un-marked.) When the checking process was done the drawing is given back to the original Draftsman to do make any corrections. This process is as important as the design and drawing. This is not some quick review this is a time consuming review of the design itself. If this step is bypassed or ignored the resulting costs for a missed error are 10 fold! (Just think of the cost of a bad titanium part?)

It has always been a rule "Measure twice, Cut once." A Draftsman knows and appreciated this, learning many lessons from the Checkers input. Murphy's law was the Draftsman's arch enemy!

The Draftsman learned from every job. Soon he/she became very knowledgeable in the standards of their industry. They became the designers of the products.

Every large Manufacturing company has a Drafting group. Even though the Drafting group was (once) part of Engineering it was basically separate with its own responsibility.

What is the Drafting group? (aka Documentation Services)
The Drafting group is only composed of Draftsman. Sometimes a large company, like Boeing, places new Engineers in the Drafting

group for a year to get an understanding of the industry standards. Drafting is all about standards.

The Drafting group is responsible for creating the final drawings and making sure those drawings are correct and meet universal and company standards. Much of the design is done by a one or a few draftsmen under the supervision of a lead Engineer.

Drafting is responsible for releasing completely defined and checked drawings to Manufacturing. This is a standard process that may be composed of many drawings that make up the product. The drawings are circulated for review and approval by specific Engineering groups, such as Manufacturing, materials and stress analysis.

Official Drawings/Documents all must contain a Title Block. The title block has all of the basic information of the drawing for reference and filing – (The drawing name and number and space for approval signatures.) Title Blocks also include: UOS (Unless otherwise specified) information, such as allowable/applicable tolerances, view orientation and "used on" information. When the title block is "signed off" the Engineering is complete and the Document(s) was delivered to the Document Control Group – who created the "blueprints" and delivered them to the relevant groups, like: Purchasing, Manufacturing or send the drawings out to suppliers for bids. Finally, the original documents are either stored in vaults, (Yes, actual vaults) or photographed for storage.

> **Note:** An Engineering drawing is a legal document (that is, a legal instrument), because it communicates all the needed information about "what is wanted" to the people who will expend resources turning the idea into a reality. It is thus a part of a contract; the purchase order and the drawing together, as well as any ancillary documents (Engineering change orders [ECOs], called-out specs), constitute the contract. Thus, if the resulting product is wrong, the worker or manufacturer are protected from liability as long as they have faithfully executed the instructions conveyed by the drawing. *If those instructions are wrong, it is the fault of the Engineer.* Because Manufacturing and construction are typically very expensive processes (involving large amounts of capital and payroll), the question of liability for errors has great legal implications as each party tries to blame the other and assign the wasted cost to the other's responsibility. This is the biggest reason why the conventions of Engineering drawing have evolved over the decades toward a very precise, unambiguous state.

What is Document Control?

It was basically an administrative group that takes the completed "drawing packages" and creates the prints, (as blueprints or microfiche) and delivers them to the appropriate areas making them available to all of the other relevant departments such as Purchasing, Tech Publications and other Engineering groups. This group like Drafting is associated with, but separate from Engineering. In many companies, today, this group is part of Engineering as included in the PLM system. But it only handles the drawings which are standard deliverables from Engineering. PLM (Product Life-cycle Management) handles the native CAD data as standard deliverables. Data inside Engineering and documents to deliver outside Engineer should be separate and handled by different groups.

Section 10: What is Manufacturing?

Abstract: Manufacturing is the making of goods by hand or by machine, that upon completion, the business sells to a customer. Items used in manufacture may be raw materials or component parts of a larger product. The Manufacturing usually happens on a large-scale production line of machinery and skilled labor.

Manufacturing is a very simple business; the owner buys the raw material or component parts to manufacture a finished product. To function as a business the manufacturer needs to cover costs, meet demand and make a product to supply the market.

A factory operates one of three types of Manufacturing production:
- Make-To-Stock (MTS) – A factory produces goods to stock stores and showrooms. By predicting the market for their goods, the manufacturer will plan production activity in advance. If they produce too much they may need to sell surplus at a loss and in producing too little they may miss the market and not sell enough to cover costs.
- Make-To-Order (MTO) – The producer waits for orders before Manufacturing stock. Inventory is easier to control and the owner does not need to rely as much on market demand. Customer waiting time is longer though and the manufacturer needs a constant stream of orders to keep the factory in production.
- Make-To-Assemble (MTA) – The factory produces component parts in anticipation of orders for assembly. By doing this, the manufacturer is ready to fulfil customer orders but if orders do not materialize, the producer will have a stock of unwanted parts.

What is Manufacturing?

I know that seems like a silly question, but articles from MSME's and PhD's who are so called PLM (Product Life-cycle Management) experts think they know how all of this works, it is very apparent that they have no clue. They sit in some ivory tower and just think how it should work. None of them has ever created a design or a parts list and probably never poured over a drawing understanding how the parts are made. They espouse the BOM (Bill of Materials) never knowing that a BOM was originally an architectural term. (Now it seems to be part of the lexicon of industrial/mechanical Engineering.) Sadly, the PLM folks are trying to expand their sphere of influence into Manufacturing. Luckily, there is much more common sense in Manufacturing and they will not fall for their failed solutions. Manufacturing "has" to deliver or they don't get paid.

PL or BOM?

In the usage for most (if not all) companies a BOM will have all of the internal sub-assembly and processing details in it (including the type of material issued, whether certs were required, or consumables used. The BOM indicates the routing of the part through various processes, etc.). A PL (Parts List) is a list of "top level" parts only, without material or "routing" information, and typically won't list any consumables used during the assembly process, (lubricants, adhesives, etc.) unless they are supplied as part of a repair kit.

However: Per ASME Y14.34-1996, paragraph 3 "Definitions", there is no difference between the two. It states, "Parts List" (PL): a tabulation of all parts and bulk materials, except those materials that support a process and are not retained, such as cleaning solvents and masking materials, used in the item.

> **Note:** Other terms previously used to describe a parts list were: list of materials, stock-list, and item list.

These are just two different terms for the same thing per the industry standards. But, it does not mean that there are not companies out there who have tried to define them as different things as noted by some of the earlier responses.

For Example:
- Parts List – A listing of all components used in the production of a parent item that does not reflect its structure or intermediate levels, and is not useful in time-phasing requirements based on lead time offsets.
- Bill of Material (BOM) – A structured list of the items used in making a parent assembly that reflects the actual production process in terms of timing and quantities consumed. It is constructed in conjunction with the routing, which describes the individual production steps and rates used. A BOM may optionally include information relating to back-flushing, use of alternate and optional components, tie between components and the operations that use them, and other data. BOMs are used by the MRP function to calculate component requirements when given a parent demand, and in building

product costs. (Synonym: product structure, recipe, formulation, ingredients list)

Manufacturing takes the final, released, drawing(s) and creates the required parts. When the parts are made they are inspected to the drawing(s) and delivered for assembly. Manufacturing is not part of Engineering or Drafting. Once they get the drawings they usually put them in a different format to use in different processes. Many companies have planning groups that manage the Manufacturing process. At assembly, Engineering may or may not supervise the process assuring that the assembly meets the functionality of the design. After that, Engineering will step out of the picture unless there are "Problems"!!

Section 11: What is a Revision?

Abstract: Sometimes Engineering is not present at assembly. Imagine an aircraft assembly line. The plane starts down this line. There is a part that doesn't fit or the assembly instructions are vague. They have a liaison Engineer that instantly handles the problem with a temporary fix. Nothing can hold up the assembly line. He/she will write up a rejection tag describing the problem and the fix. This rejection tag is sent to the responsible group.

The responsible group gets a rejection tag from Manufacturing and assigns it to a Draftsman. Why a Draftsman? The Drafting group is the most familiar with the design and documentation of the product. The Draftsman investigates the problem and working with the lead Engineer and creates a fix.

How Are Revisions Handled?
The original drawing can be difficult to change every time there is a small error and each change/iteration is very time consuming. Engineering needs a fast way to communicate the correction to Manufacturing. Manufacturing is happily creating what may be incorrect parts, thereby wasting time and material. It is very important to get that change to them as fast as possible. Many times Manufacturing has to be notified to stop making the parts.

What is an ADCN?
Advanced Drawing Change Notice: This is a document created on 8.5x11 sheets describing the fix and stapled on or added to the prints. An ADCN is released to Document Control and handled like initial released parts.

What is the DCN? (Formerly known as an ECO – Engineering Change Notice)
Drawing Change Notice: This occurs when the original drawing has to be changed. Sometimes the correction is too large to define as an ADCN. Also it may be done just to incorporate the outstanding ADCNs when times were slow. Drawings are stored in vaults. A Draftsman goes to the vault and checks out the original drawing. Again, a DCN is released the same as the initial release to Document Control.

All of the above process are done by Drafting. Engineers are never involved in this process except for the review and approval of the design.

Section 12: Engineers as Draftsman, My How Times Have Changed...

Abstract: At one time, Engineers considered producing drawings as a job that was for the "worker-bees." Almost as if the labor was beneath them, probably for good reason, since a company could hire several draftsmen for the price of one Engineer. Then, along came CAD. The idea was to make the drawing faster and increase the number of projects completed over any given time period. Someone got the idea that all those draftsmen could be replaced by the Engineer doing "their own drawings." Today, an Engineer may spend a disparate amount of time doing the work formerly done by a Draftsman and far less actual Engineering.

Think about the work for a second, near 90% of the drawing is cookie cutter work. (Does this require a four-year university degree? Once it only took two years of college – at most.)

Now consider that a new Engineer, when entering a company, has to be trained on whichever software that company uses. (Change companies and there is a good chance the "trained Engineer" has to be re-trained to use the "different" CAD System.) One has to wonder who is doing the cost/benefit analysis on this and what color the sky is in their world.

With Engineers touting their design and detailing skills – What chance does the poor Draftsman have? The Engineering world is truly upside down!

Many companies have tried to keep the Draftsman in the loop. (By creating parts and giving them to draftsmen to detail.) In the past, prints of the design layouts would be given to the Draftsman for detailing, so that they could see the relationships of the parts. It is the Draftsman's "fault" for accepting this limited amount of information.

However, if Engineers are willing default to do this type of menial design documentation, then more power to him/her. But if they are going to do it, the system has to change.

All of the standard processes that used to be the responsibility of Drafting have to be re-implemented. Today Engineers are doing "peer-checking." This is a very weak process. A successful company needs an experienced Engineer dedicated to the design-checking process. (Murphy's law sits on all of our shoulders. CAD does not "prevent accidents.")

In some companies, there are "all-levels" of Engineers where their designs and documentation will be reviewed by other Engineers of "all-levels." Are egos going to get in the way? The Manufacturing, material and stress Engineers are also going to have their input – the "hourly rate" for this review of work will soon exhaust any budget.

The Draftsman was a worker bee and, yes, many times spoke up when he/she saw a bad decision, but mostly went along with the program. Part of a Draftsman's job was to make an Engineer and Engineering "look good." Draftsmen virtually had no path to management, that was the reason contract Engineering was a great option for the skilled Draftsman.

Some argue that there is no place for the Draftsman in today's Engineering design process. Engineering is going through a transition and the Draftsman's viewpoint, which was the glue that held Engineering together, is not there to help mold it.

What Is the <u>Legal</u> Difference Between Engineering & Drafting? [14]

Years ago, there was a fairly bright line between the work performed by an Engineer and that of a Draftsman. Historically, a Draftsman sat at a Drafting table with pencils and provided drawings. Now, much of the Drafting is done through computers. Current computer software allows a Draftsman to use specifications and data to develop plan sheets with greater design information than in the days of hand drawing.

However, there is a legal distinction between Drafting and Engineering. In the "pure" context of usage, to be called an Engineer meant that a person has a rigorous academic background (at a minimum a 4-year college degree in a recognized "domain" – Mechanical, Electrical, Chemical...etc.) has passed a professional

---

14 http://www.axley.com/publication_article/what-is-the-difference-between-Engineering-and-Drafting/ *What Is the Difference Between Engineering and Drafting?* Published: September 3, 2013 Attorney Buck Sweeney Co-authored by Attorney Brian Mullins

state sanctioned exam, and has completed the "registration" process. However, the explosive growth of the Electronics Industry in the 1960's/1970's timeframe, generated company positions that were designated as "Engineer" – based on the fact that a 4-year college degree in a recognized career field was a knowledge requirement for hiring. These employees were not (generally) subjected to the rigors of professional registration. States allowed this "dilution" of the term, because these employees did not engage directly in design activities in which the public welfare or the safeguarding of life, health or property was concerned and involved, such as: the consultation, investigation, evaluation, planning, design or responsible supervision of construction, alteration, or operation, in connection with any public or private utilities, structures, projects, bridges, plants and buildings, machines, equipment, processes and works.

Note: that with the overall increase in consumer liability lawsuits, this may not remain the "norm" in the future. Therefore: it could be argued that with increasing litigation in the consumer products/consumer electronics markets it will not be long before some type of "registration" as a "Professional Engineer" is demanded of all employees whose "end-products" are used by the consumer. Industries that would most likely see this need first, are the "Medical Devices" companies.

A Professional Engineer is someone who is qualified to engage in the "practice of Professional Engineering," which is (normally) defined in most state statutes as: *...any professional service requiring the application of **Engineering principles and data**, in which the public welfare or the safeguarding of life, health or property is concerned and involved, such as consultation, investigation, evaluation, planning, design or responsible supervision of construction, alteration, or operation, in connection with any public or private utilities, structures, projects, bridges, plants and buildings, machines, equipment, processes and works...*

This would tend to "exclude" most so-named Engineering positions in commercial companies that do not meet the criteria associated with the traditional construction, or Civil-Engineering trades.

This definition relies upon, but (sadly) does not define, what constitutes "...Engineering principles and data..."

Drafting and drawing are not referenced in most of these strict legal definitions. However, courts have ruled that relaying information back and forth between "...office and the field..." does not constitute a "professional service, such as consultation, investigation, evaluation – despite the fact that the documents/drawings & specifications themselves have routinely been considered as "legal documents" by litigants.

There is a perverse conflict within the registration process itself: Notwithstanding any other provision, contractors, subcontractors or construction or material equipment suppliers are not required to register as "Professional," to perform or undertake those activities which historically and customarily have been performed by them in their respective trades and specialties, including, but not limited to:

- the preparation and use of drawings
- specifications or layouts within a construction firm or in construction operations
- superintending of construction
- installation and alteration of equipment
- cost estimating
- consultation with: Architects, Professional Engineers or owners concerning materials, equipment, methods and techniques, and investigations or consultation with respect to construction sites, provided all such activities are performed solely with respect to the performance of their work on buildings or with respect to supplies or materials furnished by them for buildings or structures on their appurtenances which are, or which are to be, enlarged, or materially altered in accordance with plans and specifications prepared by architects or Professional Engineers.

> **Note:** There is a "subtle creep" into the realm of "non-registered" employees, in that the current trend to "license" employees is an attempt to stem the tide of lawsuits and label responsible parties.

All final drawings involving the practice of Professional Engineering, prepared for the use of a firm, partnership or corporation *"...shall be dated and bear the signature and seal of the Professional Engineer who was in responsible charge of their preparation... "*

The terms "supervision," "direct supervision," "responsible charge," and "direction and control" mean "direct, personal, active supervision and control of the preparation of plans, drawings, documents, specifications . . ."

Interestingly enough, there have been very few instances of disciplinary actions against draftsmen who have been accused of practicing Engineering. The vast majority of the complaints against Engineers deal with the stamping of drawings when the Engineer does not properly supervise the Draftsman. The issue typically focuses on what constitutes "direct" supervision.

However, a statement on a drawing that a person is not performing "Engineering" services, does not provide assurance that a regulator will agree. Rather, one should expect that a regulator will evaluate whether the work requires the use of Engineering judgment to protect public health and safety. That is, a disclaimer or exclusion will likely not by itself eliminate an individual's liability.

Section 13: Acronyms…Bloody Acronyms!

Abstract: To the PLM expert, data management is Engineering's priority one problem. The high-end CAD programs do not push their design prowess on their own. That technology is available at a much lower costs today. So what makes CAD Software more valuable? Ah, yes, PLM, which is nothing more than data management, created by the convoluted minds of the PLM suppliers.

One of the first methods an Engineer encounters is MBD. MBD (Model Based Definition) seems to be pushed by inspection, where GD&T (Geometric Dimensioning & Tolerancing) is a "religion." The proponents are trying to minimize the Engineering documentation in the form of PMI (Product Manufacturing Information) delivering what you can only call a 3D drawing. (This truly is the worst attempt to deliver complete information to Manufacturing.) Sadly, there are no Draftsman (mainly because the "pictures" are "in color" and allow "visualization" to the inept eye.) there to protest this approach and Engineers seem to be oblivious to the lack of continuity in this process, since most Engineers do not understand the hierarchy of the "detail drawing" or the "fabrication drawing."

There are virtually no dimensions, just overly complex GD&T Feature Control. (Oh, yes, add GD&T "expert" to the CAD Engineer's required skills.) This is a perfect example of how a simple sheet metal part – that in the past would mostly be covered by UOS (Unless Otherwise Specified) standard industry tolerances – has been turned into an "Engineering nightmare."

Redefining 2D/3D: The "Embedded Title Block!" (A PLM Solution)

Many large Corporations bought the "Embedded Title Block" concept hook, line and sinker and it has been shown to be a horror show requiring Band-Aid after Band-Aid trying to make it work.

Today there is no "standard deliverable" that can be given to Manufacturing that equals the simplicity of what was once called a "standard drawing." Manufacturing has to jump through many differently defined hoops for each large customer (because there are no "standardized" – only company customized PLM systems) trying to complexly automate some very simple steps.

At one time, a simple but very complete drawing, that took no special software to view and understand was provided to the Manufacturing floor for fabrication/assembly and the employees "took it from there." Now, incredibly complex CAD programs and viewers are required just to "look-at" the design. (Remember, these so-called "Integrated" or 3rd party PLM and PDM data management programs are managed from inside Engineering Departments. MBE's (Model Based Engineering) complex idiosyncratic requirements, defined by groups outside Engineering however do not allow for validation programs that basically compare apples to apples. (It truly is a mess.)

Can Engineering get back to a standard process equal to what was done in the past?

There are many vested interests that would suffer if any effort to do this was attempted. Is there even the applicable knowledge of the past process to understand what needs to be done? (CAD has been here for over 40 years.) How can anyone, in today's technically evolved environment, be expected to design a "better documentation system" that would equal the standard Drafting system that was developed over centuries?

Now small companies have not fallen for this PLM/MBE fiasco and never will. This process is very costly requiring expensive software solutions and very specific expertise to maintain it. The solid model and a complete AID (Associated Information Document) as a model-based drawing, is the deliverable that is necessary for a completed Engineering design to be delivered to Manufacturing. Engineering should again take charge. A completely dimensioned detail/fabrication drawing not only adds clarity but gives the designer a second look at the design to find errors, (or even a better design.) It also provides an easy path to reviewing and a checking format that requires no special software or even a computer. Paper is very cheap, easy to handle and recyclable.

"Why MBE, MBD and PMI Will Ultimately Fail"

[MBE (Model Based Engineering)/MBD (Model Based Definition)/PMI (Product Management of Information)]

"...The big problem is, any failure will be blamed on the responsible Engineers and not an unworkable system. MBE is already being backstopped by drawings in many organizations that are forced to use MBE, but the drawings are frequently not in the release control process because they are not the 'primary' data driving fabrication. A fine mess..."

> **Note:** MBE – Model Based Engineering (MBE) is an emerging approach to engineering that holds great promise for addressing the increasing complexity of systems, and systems of systems, while reducing the time, cost, and risk to develop, deliver, and evolve these systems.

The standard deliverable from Engineering (to all groups) should not be a "native data file" requiring interpretation. It should be a PDF that includes BOTH the solid model and an AID (model based drawing) or other necessary documentation that is written directly from the CAD system file **in a standard format defined by the industry**.

Today, most programs ONLY deliver a 3D PDF. At present, disparate groups require specially trained staff to download, interpret and "re-format" the information provided by Engineering to meet their individual needs. Instead of simplifying the knowledge exchange process, the use of specialized CAD system software has added a level of complexity that has erased the primary goal of Engineering Documentation – to describe the handling, form, fit, function and architecture of a technical product or a product under development or use.

Engineers should remember the intended recipients for product documentation are both the (proficient) end-user as well as the less technically familiar administrator. In contrast to a mere "cookbook" manual, documentation aims at providing (ultimately) the service or maintenance technician enough information to understand inner and outer dependencies of the product at hand.

If technical writers are employed by the technology company, their task is to translate the usually highly formalized or abbreviated technical documentation produced during the development phase into more readable, "user-friendly" prose.

Section 14: PLM (Product Life-cycle Management) & PDM
(Product Data Management)

Abstract: Short for Product Life-cycle Management, PLM refers to a set of software tools used for mechanical design, analysis and Manufacturing to support products from when they are first conceived through distribution and retirement. PLM is assembled from various software programs, rather than purchased outright as a single commercial off-the-shelf product. Once assembled, a well-designed PLM system will manage product specifications and formulas, provide production histories, create complete product genealogies, and track total product quality.

Who needs PLM?

It is not clearly defined who needs this, but it is thought to be a requirement for large multi-year design projects, designed and manufactured in different locations, with multiple suppliers and many concurrent designers. (However, keep in mind that "multiple suppliers and many concurrent designers" also use a plethora of CAD tools: Catia, Siemens NX (was UG) and Creo (Pro/E) (High-end CAD), Solid Edge, Solidworks and Inventor – each CAD Platform offers a unique solution based on their software – and, usually, can't communicate these solutions to each other without 3rd party intervention.).

To manage a PLM system, you need a large force of IT employees and considerable infrastructure. For example, Boeing is losing 25 million on each 787 sold. Can they blame PLM alone? (It is a big part of it.) Boeing continues making it suppliers jump through hoop after hoop trying to make their (own) PLM/MBD system work with companies or suppliers who have no background, knowledge or desire to be overwhelmed by the PLM/MBD tidal wave.

For example: The 747 is a product that fits the definition of a product that would need PLM. The Airbus A-380 is a comparable product that used PLM in the design and manufacture.

There are also a multitude of stand-alone PLM products that (again) do not communicate their knowledge/data to each other without 3rd party intervention.

There are well-documented stories of errors in Engineering and Manufacturing on those projects and there are similar stories being told on some current projects. These "errors" don't cost millions, they cost billions in missed schedules and rejected parts. The problem in the former situation: Using two disparate versions of the same CAD product, Catia 4 and Catia 5 – the software releases were not compatible (backwards and forwards) and the hardware could not support BOTH outputs.

Part Numbers: another area of confusion/The Secret of Part Numbers

What was done before PLM? Before PLM there were actual detail or fabrication drawings created to go along with assembly drawings. The assembly drawings had the parts list, specifications and all the data necessary to manufacture the product.

As an explanation, to further understand the "incompatibility factor" involved, consider the example of a vendor who provides drawings of solid parts from a large Catia user in the area. The vendor is in a bit of a donnybrook with the large Catia user, trying to get clear definition of the parts. The large Catia user sees no problems in the 3D MBD, yes (3D Model Based Definition) and PMI (Product Manufacturing Information) that they are providing. (This has been an on-going struggle for this company since the project started a couple of years ago.) They even have a compatible Catia station, of course no one is trained (6 to 12-month learning curve) on the software and they use it to only get the part models to their system – Iron CAD. In the past, they have had a hard time getting the PMI, which is basically annotation GD&T in 3D space. This is not easy to view, depending on the complexity of the part, in fact, it can be a horror show and impossible to interpret. More than likely the "solution" to this problem will be the Design Engineer travelling from one company to the other in an attempt to reinterpret what was provided in the first place. There is no Universal 3D CAD today!

Hybrid Modeling and PLM: Free PMI Reader?

With Siemens NX (UG) and Pro/E bringing out hybrid modeling modules incorporated into their software, they will be able to import models into the CAD package and use their (own proprietary) PLM to maintain the parts and assembles and even modify them. So now it will not matter what package the parts are created. This is truly the next step in CAD compatibility. All packages will (eventually) have this modeling capability or they will not be in business.

(Update 8-15-16: Sadly, this function is not readily available in the design process and has not delivered the potential to standardize 3D CAD modeling.)

What is PDM? (Product Data Management)

Product Data Management is a category of computer software used to control data related to products. PDM creates and manages relations between sets of data that define a product, and store those relationships in a database. It is an important tool in Product Life-cycle Management. The "Document Control System" contained in most PLMs has so many Band-Aids on top of Band-Aids with an incredible lack of applicable knowledge of those that are involved assuring that it can never be salvaged.

A reasonable approach would be a common Web page deliverable for each part or assembly. This would be done by a separate Document Control Group, utilizing a standard release package or bundle from Engineering. This is nothing more that modifying a template

The Embedded Title Block - A PLM Solution!

Product Data Management is such a misnomer.

Draftsman never called their drawings data. "Data" is an IT term. Somehow, the concept of documentation has become skewed to mean creating "Data" to be managed by the Document Control Group. This misunderstanding of what Engineers, Designers & Draftsmen produce and deliver is what has created the convoluted PLM and MBD and its ugly stepchild the infamous PMI. Instead, it should be realized that what the ultimate reality of Drafting is – is "Part Document Management," (The basic building block of any "Product" is its individual parts.)

Those employees who have never designed a part should not be telling Engineering how to maintain assembly documentation and the associated revision process. (Realize all CAD did was add the 3D model.) Engineering has to get back in control of Engineering, how they let PLM get in charge shows an incredible lack of and a misunderstanding of the standard processes of the past. Engineers should be delivering "standard" documentation and pass it to a group that maintains and distributes the documents outside of Engineering.

There are two data streams in today's Engineering process – Native CAD and the documentation for the outside world. The CAD systems are very difficult to learn and do not lend to casual use. There is a need for a standard document, including the 3D CAD information, packaged in such a way that it can be easily duplicated with no special software or training. Today, PLM and MBD have created an industry of 3rd party products that are required to view the native CAD PMI deliverable. Sadly, there is not enough applicable knowledge to straighten this process out.

The problem with PLM/PDM is that both are trying to use the native CAD files as some sort of a "deliverable" to those parties outside of Engineering. Engineering should have should produce a standard deliverable that can be directly utilized by all of those that need the information: Purchasing and Manufacturing and depending on the type of information to: Marketing and Tech Pubs.

Most clients are already using some form of PDM even if it is just Windows Explorer. (And most of these clients are still smart enough to use drawings.) The PDF's generated from the drawing of the part or assembly is used for quoting, scheduling, Manufacturing, procuring, checking, reviewing, planning, refining, cost effect evaluation, new product definition, etc. This PDF can include all the data necessary to completely design this part or assembly. No need to have a special viewer, since the Adobe reader is available to all. You can even include a function to view in 3D. The non-Engineering folks really don't need the 3D-MBD (3D Model Based Definition), they truly only need a drawing and its specifications.

There is, also, a need to deliver a 3D model of the part to Manufacturing that should match the drawing. This puts a lot of responsibility on the design Engineering group to make sure the drawings are kept current, but this is no different from the past. Many of the drawings started out as assembly layouts, from which the drawings were created. But for some reason, currently this is considered very time consuming expense until the parts start coming in wrong or mistakes are caught by Manufacturing. Project cost over runs, delays, redesign, etc. are the result of not measuring twice and cutting once.

Part Document Management (PDM) is focused on capturing and maintaining information on products and/or services throughout the product development and useful life. Typical information managed in a PDM module includes:
- Part number
- Part description
- Supplier/vendor
- Vendor part number and description
- Unit(s) of measure
- Cost/price
- Schematic or CAD drawing
- Material data sheets

PDM Advantages: (assumed)
- Tracks and manages all changes to product related data
- Accelerates return on investment with easy setup
- Spend less time organizing and tracking design data
- Improve productivity through reuse of product design data
- Enhance collaboration within the Organization

The Drawing:
In the past the drawing was the total PDM package. (It seems as if it was a mystery how products were made.) All the above PDM functions were contained within one document that had signature blocks for:

Drawn by:	The name of the drafter/designer
Engineer:	The responsible Engineer
Stress:	The stress Engineer
Mfg. Eng.:	The Manufacturing Engineer
Eng. Mgr.:	The responsible Engineering manager
BOM	All material and size and parts if assembly.
Rev. Block:	Kept track of all revision to the part.

There was, also, an area designated on the face of the drawing for Notes: where all specs unique to the materials or Assembly/rigging instructions were written an annotated.

Yes, one document, compare that to now. A documentation "user" must have a special software to view the PMI. 3D-MBD has dimensions all over, to the point where they are too confusing to review. The features that are covered by the profile tolerance are not dimensioned, leaving them to actually be overlooked by the designer. There are parts that could not have gone thru all the above checks. However, the false sense of "control" fostered by "Data management" does not guarantee these parts can be manufactured and many parts do not fit together at assembly. In reality, there are no short cuts to complete Engineering procedure.

Section 15: Enter the 3D Model:

Abstract: Many think the 3D model does everything. Sad to say it really doesn't. Yes, we can use the model for new products. But to manufacture the part, it has to be documented. Many of the parts you will review have been created by CAD jockeys and from the looks of them not reviewed for manufacturability. These parts have been rejected time and time again by the suppliers, or many times they have been made – only not to fit in the assembly. Engineering is a time consuming endeavor. It has to be done as correct as it can be done. It has to have checks, and rechecks because "MURPHY'S LAW" is in full effect. If you supply the drawing and the solid model in the same package that is verified by the responsible design group, you will be assured that the correct part is being manufactured. From the looks of the results, the 3D model is becoming more of a hindrance then an asset, at least in the current form of delivery.

Learning Mechanical CAD

This is really 3D Drafting. It serves the exact same purpose of the process done in the past. We have the same design challenges, nothing has changed. We still have to put screws in holes and make sure parts will fit. Nothing has changed except that now – a degreed Engineer is now doing the same job (at a much higher hourly rate) that a skilled draftsman once did. Engineers now design non-critical products components that do not require any of their analysis skills. They have become the low-end designers, known in the past as the "Draftsman". Sadly, they will not know any better. (But the Engineering management path, that many young Engineers dreamed of, has just become much more narrow.)

The introduction of the 3D model, as the basis for design, has changed the process completely. By eliminating the drawing or layout in the beginning, and now using 3D CAD, product design has become a single step – with the documentation created at the end of the process, instead of at the beginning of the process, all being performed by a single designer. The skills of the designer with the orthographic drawing or layout are not needed and some say we must face the cold hard facts: The Draftsman is not needed.

The Death of the Drawing

But per the explanation above we still need Drafting. Do we need a new profession or will the degreed Engineer be happy providing this level of design? Certainly some Engineers seem to be happy to do the simple grunt work of form, fit and function of the design that was left to the Draftsman in the past. (But this type of design is not taught in college.) I wonder if there will be a smooth process of learning this on the job as it was done in the past with the Draftsman. CAD will only be taught in college, in the future, as soon as the type of CAD system becomes irrelevant. As of today, CAD systems represent the latest "Tower of Babel" …

Section 16: Evaluating Your Design

Abstract: After completing (what you believe to be) a "Great Design" there is a period of time to be reserved for Evaluating Your design and most likely, changing certain elements given the following considerations. **Note:** Through experience you will "automatically" think of your design in this way and eventually (just before you retire), you will no longer need to perform these evaluations outside your "Design-Time"

Evaluation Based on Technology-Readiness Assessment

- Absolute comparison with state-of-the-art capabilities. (Matured Technology)
- Six measures to determine a technology maturity
  a) Can the technology be manufactured with known processes?
  b) Are the critical parameters (such as dimensions, material properties, or other features) that control the function identified?
  c) Are the safe operating latitude (limits) and sensitivity of the parameters known?
  d) Have the failure modes been identified?
  e) Does hardware exist that demonstrates positive answers to the above 4 questions?
  f) Is the technology controllable throughout the products life cycle?

Evaluation for Design Robustness:

The word "robust" in design usually refers to final products that are of high quality because they are insensitive to Manufacturing variation, operating temperature, wear, and other uncontrolled factors so that performance is maintained.

Evaluation for Design Reliability: Failure Mode & Effects Analysis Failure mode and effects analysis (FMEA) was developed in the 1950s and is used to evaluate designs at their early stages in terms of reliability. The criteria are also, very useful to highlight both the need for and the effects of design changes. The method involves listing all possible failure modes for each component with their effects on the device subsystems.

FMEA requires that the following steps be performed:

- Define the boundaries of the system under consideration and its associated detailed requirements.
- List all system components and subsystems.
- Identify, in writing, and list each component's failure modes, including a clear description.
- Assign a failure rate or failure probability to each component failure mode.
    - List each failure mode effect on the subsystem, system, and plant.
    - Enter remarks for each identified failure mode.
- Review critical failure modes and take appropriate corrective measures.

There are many benefits of performing FMEA: the analysis provides a systematic approach to classify hardware failures, lower development time and cost, reduce Engineering changes, and is easy to understand. FMEA serves as a useful tool for more efficient test planning and highlights safety concerns. Furthermore, this method can improve customer satisfaction and serve as an effective tool to analyze small, large, and complex systems.

Perhaps most importantly, FMEA provides a safeguard against repeating the same mistakes in the future and improves communication among design interface personnel. The application of this method during the initial stages of medical device design can be very useful.

Evaluation: Design for Manufacturing [15]
A number of general design guidelines have been established to achieve higher quality, lower cost, improved application of automation and better maintainability. Examples of these DFM guidelines are as follows:
- Reduce the number of parts to minimize the opportunity for a defective part or an assembly error, to decrease the total cost of fabricating and assembling the product, and to improve the chance to automate the process
- Foolproof the assembly design (poka-yoke) so that the assembly process is unambiguous

---

[15] http://www.npd-solutions.com/dfm.html

- Design verifiability into the product and its components to provide a natural test or inspection of the item
- Avoid tight tolerances beyond the natural capability of the Manufacturing processes and design in the middle of a part's tolerance range
- Design "robustness" into products to compensate for uncertainty in the product's Manufacturing, testing and use
- Design for parts orientation and handling to minimize non-value-added manual effort, to avoid ambiguity in orienting and merging parts, and to facilitate automation
- Design for ease of assembly by utilizing simple patterns of movement and minimizing fastening steps
- Utilize common parts and materials to facilitate design activities, to minimize the amount of inventory in the system and to standardize handling and assembly operations
- Design modular products to facilitate assembly with building block components and sub-assemblies
- Design for ease of servicing the product

Design for Manufacturability and Integrated Product Development may require additional effort early in the design process. However, the integration of product and process design through improved business practices, management philosophies and technology tools will result in a more producible product to better meet customer needs, a quicker and smoother transition to Manufacturing, and a lower total program/life cycle cost.

In an increasingly competitive world, product design and customer service may be the ultimate way to distinguish a company's capabilities. Because of the growing importance of product design, Design for Manufacturability and Integrated Product Development concepts will be critical. It will be the key to achieving and sustaining competitive advantage through the development of high quality, highly functional products effectively manufactured through the synergy of integrated product and process design.

Evaluation: Design for Assembly [16]

The aim of design for assembly (DFA) is to simplify the product so that the cost of assembly is reduced. However, consequences of applying DFA usually include improved quality and reliability, and a reduction in production equipment and part inventory. These secondary benefits often outweigh the cost reductions in assembly. DFA recognizes the need to analyses both the part design and the whole product for any assembly problems early in the design process. We may define DFA as: "…a process for improving product design for easy and low-cost assembly, focusing on functionality and on assemblability concurrently."

The practice of DFA as a distinct feature of designing is a relatively recent development, but many companies have been essentially doing DFA for a long time. For example, General Electric published an internal Manufacturing Producibility handbook in the 1960's as a set of guidelines and Manufacturing data for designers to follow. These guidelines embedded many of the principles of DFA without ever actually calling it that or distinguishing it from the rest of the product development process.

It wasn't until the 1970's that papers and books on the topic began to appear. Most important among these were the publications of G. Boothroyd that promoted the use of DFA in industry.

Basic DFA Guidelines

Here are some basic guidelines for DFA. Generally, you want to start with a concept design and then go through each of these guidelines, decide whether or not it is applicable, and the modify the concept to satisfy the guideline. There is no guarantee that a given guideline will apply to a particular design problem. Many of these guidelines are similar or the same as rules of concurrent Engineering:

- Minimize part count by incorporating multiple functions into single parts
- Modularize multiple parts into single subassemblies
- Assemble in open space, not in confined spaces; never bury important components

---

16 Design for Assembly: Vincent Chan and Filippo A. Salustri

- Make parts such that it is easy to identify how they should be oriented for insertion
- Prefer self-locating parts
- Standardize to reduce part variety
- Maximize part symmetry
- Design in geometric or weight polar properties if nonsymmetrical
- Eliminate "tangly" parts
- Color-code parts that are different but shaped similarly
- Prevent nesting of parts; prefer stacked assemblies
- Provide orienting features on non-symmetries
- Design the mating features for easy insertion
- Provide alignment features
- Insert new parts into an assembly from above
- Eliminate re-orientation of both parts and assemblies
- Eliminate fasteners
- Place fasteners away from obstructions; design in fastener access
- Deep channels should be sufficiently wide to provide access to fastening tools; eliminate channels if possible
- Provide flats for uniform fastening and fastening ease
- Ensure sufficient space between fasteners and other features for a fastening tool
- Prefer easily handled parts

Design Guidelines for Hard Automation

The main different here is that assembly is performed by machines instead of by humans.

- Reduce the number of different components by considering does the part move relative to other parts?
- Must the part be isolated from other parts (electrical, vibration, etc.)?
- Must the part be separate to allow assembly (cover plates, etc.)?
- Use self-aligning and self-locating features
- Avoid screws/bolts
- Use the largest and most rigid part as the assembly base and fixture.
- Assembly should be performed in a layered, bottom-up manner.

- Use standard components and materials.
- Avoid tangling or nesting parts.
- Avoid flexible and fragile parts.
- Avoid parts that require orientation.
- Use parts that can be fed automatically.
- Design parts with a low center of gravity.

Sometimes it is too difficult to make parts symmetrical, often non-functional features are added to a part to facilitate part feeding, grasping, and orientation.

Design Guidelines for Manual Assembly

Obviously, the following guidelines depend on the skill of the worker:

- Eliminate the need for workers to make decisions or adjustments.
- Ensure accessibility and visibility.
- Eliminate the need for assembly tools and gauges (i.e. Prefer self-locating parts).
- Minimize the number of different parts - use "standard" parts.
- Minimize the number of parts.
- Avoid or minimize part orientation during assembly (i.e. Prefer symmetrical parts).
- Prefer easily handled parts that do not tangle or nest within one another.

Note that many products do not lend themselves to these guidelines. Many such products are sold as "ready-to-assemble" kits or require that assembly be shifted to cheaper labor markets.

"Robustness" also refers to decisions that are as insensitive as possible, to the uncertainty, incompleteness, and evolution of the information that they are based on.

Satisfaction = belief that an alternative meets the criteria

Belief = knowledge + confidence

Section 17: A Word About Purchasing

Abstract: Purchasing is the department that delivers the drawings to Manufacturing, to in-house or outside suppliers for quotes. Purchasing keeps track of the document revisions and where standardized parts and components can be used across product lines. The Purchasing drawings/specifications should include "used on" information.

For example, the Purchasing documents in large corporations would have the different effectivities (examples: "blocks" of airplanes or "models" of vehicles) for the different assemblies used. Many times, one Purchasing drawing/specification would have many different configurations for different aircraft types (defined by dash numbers) or vehicle configurations. This is another place where PLM (Product Life-cycle Management) has failed. Due to the way CAD (Reference: The Pro/e paradigm) is set up, Purchasing must handle referenced, external parts, in a special fashion. While this may be advantageous for the conceptual design, it is a horror show for final released projects and items to be used as deliverables. The information should all be in one single file where all information is available without resorting to a convoluted native file system.

Epilog [17]

President Herbert Hoover used to tell of meeting a woman on a ship while he was traveling. After several conversations over a week or so, the woman asked what his occupation was. Hoover told her he was an engineer, a Mining Engineer. The woman replied, "An engineer? I thought you were a gentleman." It seems the lady, like many people of her time, assumed engineering was not a gentlemanly career.

This lack of respect from society still galls many in engineering, especially when compared to the public adulation (or at least high salaries) given to doctors, lawyers, and scientists.

"...Part of the respect problem is that the personality of the average engineer and the way they are taught does not bring out the best as far as earning respect from society..." says Henry Petroski, a professor of Civil Engineering and History at Duke University, and author of several books on engineering.

Another reason Engineers may not get the respect they deserve is that they always seem to be working in the shadows of scientists and leaving the public unsure of what engineers really do, (besides drive trains.) *"...The public just isn't too savvy. They confuse science and engineering, almost always to the detriment of engineers..."* says Edward Pershey, vice president of special projects at the Western Reserve Historical Society with a Ph.D. in the History of Technology. *"...A scientist's goal is to uncover new information about how the world works. Engineers take this knowledge and solve problems..."*

(Engineers in the early days of the space race used to tell the story that when a rocket launched successfully, it was called a scientific breakthrough. But if it exploded on the pad, or shortly thereafter, it was called an "engineering failure.")

One move engineers took back in the mid-1800s to increase their stature in society, was to form professional organizations. Eventually, this grew into the Professional Engineers license in the early to mid-

---

17 Changes in the Engineering Profession Over 80 Years, Apr 7, 2009 Stephen Mraz
http://machinedesign.com/technologies/changes-engineering-profession-over-80-years

1900s – a way to ensure only those educated and trained in orthodox engineering could call themselves engineers. *"...It was a way to regulate the profession and to give more respect to 'real' engineers..."* says Petroski.

With some states, it took a public disaster and great loss of life before anything was done to ensure only qualified people could legally call themselves engineers. In Texas, for example, the New London public school was destroyed when its gas-fired boiler exploded. At least 295 students and teachers were killed and many others left injured and maimed. This tragedy spurred Texas to pass a registration law requiring those wanting to call themselves engineers to meet certain requirements and become licensed. (The accident also led to the requirement that Ethyl-Mercaptan, a substance that smells like rotten eggs, be added to natural gas to give the odorless fuel a telltale smell, one that would alert most people to a gas leak.)

"...So now each state has its own definition of what it takes to be an engineer, which is similar to doctors and lawyers who must be licensed to practice in a particular state..." (Petroski)

Needless to say, PE (Professional Engineer) licensing has not garnered much respect for engineers over the last half century. This might be due to companies like GE, Lockheed and Boeing that hire lots of engineers, (actually college graduates of Engineering Degree Programs) but haven't been too keen on PE licensing. *"...Companies didn't want to lose control of their employees and have them form unions..."* says Petroski. *"...They might have thought those with PE licenses would want more money and perhaps more say in how they did their jobs. As a result of this pressure from companies and no organization on engineers' part, there has been little legislation requiring a PE license for engineering jobs and projects..."*

One aspect of the profession that has not changed, is the basic job of engineering: To solve problems economically. Or as Petroski puts it, *"...to carry out the engineering method and pursue institutionalized invention..."*

"...There will always be new tools and new technologies. But what goes on in an engineer's head isn't going to change too much..."

An Electro-Mechanical Engineer's primary responsibilities will usually fall under two headings: (either) the sustaining engineering support for existing products, OR the development of Electro-Mechanical Packaging conceptual designs for new products, to include testing and validation of new and redesigned products. Both descriptions include Manufacturing and Sales support.

Conceptual Design – Develop conceptual Electro-Mechanical Packaging designs for new Electro-Mechanical and Mechanical assemblies and sub-assemblies to support new or existing product redesign initiatives. This includes the development of Engineering Product Requirements and documenting Specifications, based on initial functional and end use review.

Detailed Design – Develop detailed Mechanical designs to support established design requirements using sate of the art Mechanical design software tools. This includes user interfaces, Electro-Mechanical interfaces, material specification(s), machining and assembly requirements.

Manufacturing Engineering – Complete required activities for new product and design change product releases. This includes industry standard prints/models and required change documentation such as BOM's, prints, product manuals and MRP/ERP changes as required

Research & Development – Research the application of new or "new to the company" technologies for use in new incremental, platform or breakthrough product development. Assess the feasibility and cost of implementing these new technologies to produce cost effective new products. Direct the Research & Development Machine Shop in the fabrication of components for research activities including new product design validation prototypes, finished product validation and new technology implementation.

Product Component Testing and Validation – The testing and validation of new and redesigned mechanical components and mechanical support of finished product testing. Testing to validate changes for mechanical issues found post-product release.

<u>Customer Support</u> – Provide customer support for custom orders and installations and be willing to travel to customer sites for installation of products as needed.

Additional duties and responsibilities as assigned by the Engineering Director.

Qualifications of a typical Electro-Mechanical Engineer include:
A Bachelor's Degree in Mechanical Engineering (or a related discipline.)

- Three-to-five years' experience as a Mechanical Engineer in a Manufacturing or Engineering design environment – Preferably in an Electro-Mechanical design & Manufacturing environment.
- Motivated team player with excellent communication skills, ability to work directly with people of varying backgrounds, have a strong technical foundation, superior attention to detail, and be able to work independently and or with a cross functional group.
- Experience managing a complete product development cycle from initial design specifications through complete Manufacturing documentation.
- Experience with Product Life-cycle Management (PLM), PLM software systems and phase gate development is a plus.
- Experience with Quality Systems (Corrective Action, APQP, etc.) is preferred.
- Forward-thinking and improvement-minded experience in a product development team environment is required.
- Team leadership skills in an engineering or technical environment is a plus.
- Technical skills including:
  - Mechanical and Electro-Mechanical design
  - Material specification
  - Detailed Engineering & Manufacturing Documentation.
- A proficiency with 3D Mechanical CAD software and the ability to create fabrication documents to industry standards.
- Experience with Solid Works and Model Based Engineering (MBE) is a plus.

Acknowledgements

(also in footnotes)

Wikipedia: a resource used to verify the validity of the information presented.

System Design Theory: A Process, First Published:27 July 1981, John L Bisol

Evaluation Procedure for Cabinet/Enclosure Design, First Published:19 Sept. 1983, John L Bisol; Presented:16 Nov. 1983 @ Nepcon Central

Methods for Prediction of Electro-Mechanical System Reliability, by Thomas L. Bush, Anthony P. Meyers, & Darwin F. Simonaitis. Published in the Semi-annual report for the U.S. Army Electronics Laboratory @ Ft. Monmouth N.J., 15 May 1964 to 14 November 1964; pages 46 and 47.

Product Management: Eight Simple Steps for New Product Development http//www.business2community.com/product-management/eight-simple-steps-for-new-product-development-0560298#RB1rIhUgWD5CxagZ.99   10/19/2016

NASA Tech Brief, WINTER 1980 Vol. 5, No. 4. MSC-18745

Effects of Product Failure Severity and Locus of Causality on Consumers' Brand Evaluation: Song, Sujin; Sheinin, Dan A.; Yoon, Sukki: Social Behavior and Personality: an international journal, Volume 44, Number 7, 2016, pp. 1209-1221(13) Scientific Journal Publishers

America Goes Green: An Encyclopedia of Eco-Friendly Culture in the United States; Volume I: Thematic Entries, Kim Kennedy White, ABC – CLIO, LLC., 2003, Page 675

MIL-HDBK-217F: Reliability Prediction of Electronic Equipment
02 December 1991 – Superseding: MIL-HDBK-217E, Notice 1-12 January 1990

MIL-HDBK-338B: Electronic Reliability Design Handbook
1 October 1998 – Superseding: MIL-HDBK-338A

5 Unbelievable Product Liability Lawsuits – April 8, 2013 Comments in Courtroom by Lisa; http://www.iveyEngineering.com/unbelievable-product-liability-lawsuits/ 16/10/2016

How Much Is Your Arm Worth Depends On Where You Work: by Michael Grabell, ProPublica, & Howard Berkes, NPR March 5, 2015   10/19/2016 https://www.propublica.org/article/how-much-is-your-arm-worth-depends-where-you-work

Why Those Samsung Batteries Exploded, http://www.chicagotribune.com/bluesky/technology/ct-why-those-samsung-batteries-exploded-wp-bsi-20160912-story.html   10/15/2016

Thermomix Products Responsible for 45 Injuries says Consumer Group: CHOICE https://www.theguardian.com/australia-news/2016/may/12/thermomix-products-reponsible-for-45-injuries-says-consumer-group-choice   10/18/2016

What Is the Difference Between Engineering and Drafting? Published: September 3, 2013, by Attorney Buck Sweeney Co-authored by Attorney Brian Mullins http://www.axley.com/publication_article/what-is-the-difference-between - Engineering-and-Drafting/ 10/19/2016

Evaluation: Design for Manufacturing; http://www.npd-solutions.com/dfm.html 10/19/2016

Design for Assembly; Vincent Chan and Filippo A. Salustri

The Death of the Draftsman or "Where Has All the Talent Gone?" by Joe Brouwer; Tech-Net, Inc., http://www.tecnetinc.com/index.htm 10/20/2016

Changes in the Engineering Profession Over 80 Years, Apr 7, 2009 Stephen Mraz http://machinedesign.com/technologies/changes-engineering-profession-over-80-years 10/20/2016

Numerous personal files, personal documents and previously published technical articles created or owned by the author are included – either wholly or in part – as the intellectual input for this book.

Other Books by John L. Bisol

1. The House on South Street: second Edition   ISBN: 978-1-365-04162-4
2. Rendition I   ISBN: 978-1-4116-2220-3
3. Rendition II   ISBN: 978-0-6151-3560-1
4. BISOL – The Knight Templar   ISBN: 978-1-8472-8923-0
5. SURVI V0R – The House Sale   ISBN: 978-1-312-98902-3
6. Out of Sight – Out of Mind   ISBN: 978-1-312-94704-7
7. The House On Ernes Drive   ISBN: 978-1-312-91172-2
8. An Ostentation of Tutoring   ISBN: 978-1-329-64084-9
9. The Worcester Tornado: A New Perspective   ISBN: 978-1-329-70645-3
10. The Golden Alders   ISBN: 978-1-329-00035-3
11. The House on Middle Street   ISBN: 978-1-329-77449-0
12. The House on South Street: Revisited   ISBN: 978-1-329-83038-7
13. Occupational Education: Insights & Perspectives   ISBN: 978-1-329-90631-0
14. The Veil of Cadence Shadowsoul   ISBN: 978-1-329-94140-3
15. Cleaning the House on South Street   ISBN: 978-1-365-23100-1
16. The Stories of: The House on South Street   ISBN: 978-1-365-23100-1
17. The Third Rendition   ISBN: 978-1-365-31587-9
18. Me & My Conditions   ISBN: 978-1-365-38142-3
19. The House on Nichols Street   ISBN: 978-1-365-38272-7
20. An Interpretation of the Artworks: St. Francis of Assisi Church – Fitchburg, MA   ISBN: 978-1-329-71156-3

How to Order:

All books are available on: amazon.com OR by ISBN @ Your Bookstore
ALSO:
http://www.lulu.com/spotlight/jlbisol
OR:
http://www.amazon.com/John-L.-Bisol/e/B00US30LJE

Special Order Book - OBLIGATIONS: A One-Act Play
(contact Author for License) @ www.bisol.net

www.ingramcontent.com/pod-product-compliance
Lightning Source LLC
Chambersburg PA
CBHW021950170526
45157CB00003B/930